D0119218

THE LONGEVITY SEEKERS

Science, Business, and the Fountain of Youth

TED ANTON

THE UNIVERSITY OF CHICAGO PRESS CHICAGO AND LONDON

Ted Anton is professor of English at DePaul University. He is the author, most recently, of *Bold Science* and has written for *Chicago* magazine, the *Chicago Tribune*, and *Publishers Weekly*.

The University of Chicago Press, Chicago 60637
The University of Chicago Press, Ltd., London

Printed in the United States of America

22 21 20 19 18 17 16 15 14 13   1 2 3 4 5

ISBN-13: 978-0-226-02093-8 (cloth)
ISBN-13: 978-0-226-02095-2 (e-book)

Library of Congress Cataloging-in-Publication Data

Anton, Ted.
The longevity seekers : science, business, and the fountain of youth / Ted Anton.
pages. cm.—(From obscurity, 1980–2005—"Greater than the double helix itself," 1980–1990—The grim reaper, 1991–1993—Sorcerer's apprentices, 1991–1996—Race for a master switch, 1989–2000—Money to burn, 2000–2003—Longevity noir, 2003–2004—Betting the trifecta, 2005–2006—Defying gravity: the battle to find a drug for extending health, 2005–2013—Sex, power and the wild: the evolution of aging, 2001–2008—The rush and crisis, 2008–2010—Live long and prosper, 2009–2011—Centenarians in the making, 2011–2013—Fountains of youth, 2013—Reimagining age.)
ISBN 978-0-226-02093-8 (alk. paper)—ISBN 978-0-226-02095-2 (e-book) 1. Life expectancy—Economic aspects. 2. Longevity—Economic aspects. 3. Life spans (Biology) 4. Old age—Economic aspects. I. Title.
HB1322.3.A58 2013
612.6′8—dc23
2012043340

♾ This paper meets the requirements of ANSI/NISO Z39.48–1992 (Permanence of Paper).

For my parents, Bertha and Gus

Contents

A Note on Purpose *ix*

Preface *xi*

PART 1. FROM OBSCURITY, 1980–2005 *1*

1 "Greater Than the Double Helix Itself," 1980–1990 *3*

2 The Grim Reaper, 1991–1993 *14*

3 Sorcerer's Apprentices, 1991–1996 *23*

4 Race for a Master Switch, 1989–2000 *33*

5 Money to Burn, 2000–2003 *44*

6 Longevity Noir, 2003- 2004 *57*

7 Betting the Trifecta, 2005–2006 *73*

PART 2. DEFYING GRAVITY: THE BATTLE TO FIND A DRUG FOR EXTENDING HEALTH, 2005–2013 *85*

8 Sex, Power and the Wild: The Evolution of Aging, 2001–2008 *87*

9 The Rush and Crisis, 2008–2010 *100*

10 Live Long and Prosper, 2009–2011 *113*

11 Centenarians in the Making, 2011–2013 *128*

12 Fountains of Youth, 2013– *141*

13 Reimagining Age *156*

Epilogue *171*

Acknowledgments *177*

Longevity Gene Timeline *179*

List of Longevity Genes *181*

Notes *183*

About the Author *213*

Index *215*

A Note on Purpose

This story began with my interest in an article in the *New York Times* science section. In January 2001, I called the article's subject, MIT biologist Lenny Guarente, who suggested I call the University of California at San Francisco's Cynthia Kenyon. Over the next eleven years I became hooked on their science of longevity genes. I conducted more than two hundred interviews with scientists, investors, and students in their offices and labs. I visited and worked in labs, observed classes and conferences, and traveled from California to Crete to research the story of the science behind the dream of extending healthful life.

To research the book, I attended conferences in Cold Spring Harbor, New York (2002); Hersonnissos, Crete (2004); St. Louis, Missouri (2006); Boston, Massachusetts (2008 and 2009); and Madison, Wisconsin (2010). I watched scientists teach classes, run their labs, and conduct journal seminars. I was a student for a day in the Woods Hole, Massachusetts, Genetics of Aging course.

The following pages explore the social history of a science idea. A researcher grows up in a creative family, follows an obscure interest that appeals to them and no one else, and stumbles onto an exciting insight that alters a mode of thought. I am interested in the power of such ideas, revealing why some receive attention at a given moment and others do not. This book proposes that in such behind-the-scenes moments, one can discern crucial clues to the historical, technological, and social forces driving an era.

Naturally, memories of such moments may vary. As much as possible I have sought to square my interpretation with the recollections of the scientists. I have been fortunate that the researchers themselves are recording the history of their field in professional journals, online and print interviews, conferences, and in their own books or oral histories. I consulted these works, talked to as many lab members as possible, and tried to match recollections with the record and those of other lab members. I went back to most researchers and checked the details of a scene. What follows is my attempt to recount the unfolding of an absorbing and disputed series of discoveries. If I quote a conversation, it is based on interviews or public documents. Picking certain researchers and discoveries, however, I leave others out. Ultimately, my goal is not to provide a comprehensive text, but a dramatic record of the personal, economic, and intellectual motivations shaping discovery in our time.

One of these discoveries could reshape the way we experience our lives. In the race for longer, healthier lives, almost every discovery came under explosive disagreement. What role did accident or personality play in their unfolding? To what extent did public fascination and big money alter the course of science? Which of these findings, if any, may lead to significant applications? This book explores the relation of a unique science to its time and, in so doing, the relation of any science to any time.

Preface

The Laboratory of Molecular Biology sat at the end of Hills Road on the southern edge of Cambridge, England. In 1983 the weather had been so miserable that twenty-nine-year-old Cynthia Kenyon taped a yellow sun on her single window overlooking distant hedgerows and a lone traffic light. She was checking her experiments in her tiny three feet of bench space in a room in one of the leading institutions of molecular biology. The room was small and crammed with equipment, with cream-colored walls. Upstairs was a cafeteria strewn with newspapers and ashtrays. The laboratory smelled of coffee.

She worked with spectacular people mostly in their twenties who were committed to science driven by ideas. Their main idea was that if you had a biological question, you studied it in a living animal, not in single cells. Lanky and tall, with short blond hair and freckles, Cynthia Kenyon had changed her career to join this group. The oldest of three who grew up in New Jersey and Georgia, she hated limits and disliked authority. In high school, she played jokes on the band master even though she dreamed of a concert career playing French horn. She hung a banner, "Know the truth and it shall make you free," in her bedroom, where she allowed a parakeet to fly free and taught it to pick playing cards from her hand. She wrote stories, played guitar, kept a huge aquarium, and sewed her own clothes. She yearned to do something great.

That morning the rain had finally stopped. The lab emptied as people headed out to enjoy the sunshine. An opera played on the radio. At her bench, Kenyon noticed one of the petri dishes of tiny worms. She pulled over a microscope to take a closer look. The tiny worms were called *Caenorhabditis elegans*, "elegans" for their elegant, sinuous, near-transparent bodies. *C. elegans* lives in the temperate soils and decaying fruits of the earth. It consists of almost every tissue that a human body contains, but is only the size of a comma in a printed sentence.

The worms in Kenyon's dish barely moved. Their skin bunched in weedy, menacing clumps. Their backs looked bloated and thick. They looked old and near death. They looked, she thought suddenly, like people. They aged just like humans. Then she realized she too was getting old, and someday she would die.

The mystery the lab studied was the molecular plan of growth, repeated beautifully every time, that controlled the way nerve and other cells organized in patterns to build an animal. It took tens of millions of steps to make an animal from a one-celled embryo, each choreographed by a mysterious gene program. The revolutionary discovery was that single genes, shared among species, controlled many of key steps of the growth program. But the revolution was over, and Kenyon had mostly missed it.

No one thought that a similar pattern might determine the rate of our decline. Aging was random decay, and no one wanted to study that. Yet aging was one of life's most important processes, she thought, and all biological processes, they were learning, were controlled by genes. If everything was in some way controlled, then aging might be as well.

Some moments you feel or see something that sticks in the unconscious like a dream. Outside her window the traffic light changed. She looked at the plate of worms. They barely moved. She never forgot that moment. It was a beautiful day and her heart was pounding.

This is the story of the race to understand the genes of healthful human longevity. For years, researchers have extended the life of lab animals up to ten times their normal span. Working on tiny worms, flies, fungi, and mice, scientists discovered molecules that sense nutrient and energy levels and extend fitness into late life.

The question was whether the same may be true for humans. Scientists' new insight was that the rate of aging may not be random and chaotic, but rather a controlled and perhaps manipulable process. By changing the activity of only a few genes, we may be able to live much longer, healthier lives. No science ever received quite so much public fascination because few have offered such immediate promise of potential social impact. We face an aging crisis. If current trends continue, the numbers of people older than 60 years will more than double by 2050. By that year, one in three people in the developed world will be older than sixty. The intensity of this "silver tsunami" is even greater in the developing world, where countries will have less time to adjust. Across the globe, the number of centenarians will increase eighteen-fold in the next fifty years. Even though we are living longer, we still suffer the tragedy of late-onset illness. By a large margin, most health care costs are incurred at the end of life. A host of policy makers today are arguing about what the aging world means and how to address it.

This book tells the story of a potential science revolution and of the new money changing the way scientific ideas emerge. It begins with geneticist Cynthia Kenyon, 59, who found one of the world's first longevity genes and cofounded a company to find a drug to extend healthful human life, competing with several new companies, many funded by the barons of the information age.

Rather than spending billions of dollars to battle the various diseases of aging, these researchers argued, we might better spend our efforts on a new way of thinking, at the level of molecular tipping points in the cell that make the beautiful circuits and feedback loops that can either maintain or damage our health. These tipping points come in small switches triggered by nutrients, energy levels, and other changes in the environment. These switches can almost miraculously calibrate the rate of aging in many species, from familiar lab models to humans.

This is the biography of an idea. The idea is that the quality of aging could be altered by tweaking single genes. It originated among a handful of outsiders, working on tiny animals outside the mainstreams of traditional aging study. There was no textbook or blueprint for pursuing that idea, but there was a culture of small groups often at odds with each other. The early researchers fought for money, and, if they got it from the super-rich, they worked without many of the ties of conventional funding. In the scramble for scientific credit and financial gain, the battle to survive required skills of control, cooperation, and competition for resources, much like the genes themselves.

Aging remains one of life's great unsolved riddles. For all our biological knowledge, we still do not know exactly what aging is. Life can defy entropy, the tendency of systems to wear down over time. Healthful life span may last as long as it wants or needs to. Some organisms live centuries or longer. We now know something of the gene, cell, and hormone signals that help them do so.

This book explores the mystery of those intricate cell signals and the battles of scientific personalities to expose them. The discoveries offer deep insights, but equally as illuminating was the confluence of dreams, personal crises, intense work, altruism, and greed that made one of the most contentious science stories of our time. Few recent scientific discoveries have moved from such an outlier status to the pinnacle of business, while riling up more critics. Some ideas won, and some of the biggest lost out.

We are all intrigued by the questions of life and death, the more so the older we get. What are the proper stages of life? In ancient Rome, people lived twenty-nine years. In the Middle Ages, the afterlife and current life were intimately commingled because they had to be. In the last twenty-five years a technical revolution made it possible to answer such questions as: What causes aging? Can we alter its rate? Does life do so naturally? What would be the meaning of such discoveries, and which are the ones most likely to affect the ways we live? This book explores the attempts to answer these big questions. To do so could change our definition of who we are and how we imagine ourselves.

The book's main subjects are four genes and their molecular signaling pathways. They offer a direct evolutionary gift, from ancient common default signals governing growth, reproduction and stress response, of potentially longer and healthier lives. When the implications of that gift were understood, the science exploded in the media, medicine and industry, and companies and governments invested more than a billion dollars to wrest the secrets of youth from the laboratory.

For eleven years, I followed the scientists' conferences, visited their labs, hung out in their homes, and traveled the world to understand the latest discoveries. I became obsessed by their arguments and the long, beautiful, twisting molecules that triggered them. In those years I watched as the science exploded. The backlash unleashed campaigns of mass persuasion, changing the relationships among academic, corporate, and government institutions, raising questions about what a scientist is and does, and the roles of science and science funding in our time.

This story offers a parable of science, from the contribution of idealistic undergraduate to seasoned researcher, from giant government funding agency to private billionaire, as it intersected with capital and human desire in the biggest dream of all.

Aging is universal. We live it every day. We see it in our friends and families. Extending healthful longevity made the ultimate science prize. No other science field is like it, not global warming or energy or the origins of the universe. None of those may give us a few more minutes on earth. Perhaps tweaking a worm, fly, or mouse gene could.

During these years, my father suffered a stroke and a number of people I loved have either died or shown the most debilitating effects of aging. As I entered my fifties, I felt for the first time the hint of my mortality, real as the roughness of a stone. To slow aging is to alter the program of our life. This is the story of the quest to understand and defy one of the universe's most inexorable forces. To understand it is to see the world anew.

PART 1: FROM OBSCURITY

Gilgamesh thought he would be immortal. The vase painters of ancient Athens recorded their balding selves having outlandishly youthful sex. Beneath all our exploration of longevity lies a bargain. To live longer, or better, many of us would give almost anything. Chapter 1 describes the traditional evolutionary theories explaining aging as a trade-off and explores two small 1980s discoveries in an obscure worm that seemed to suggest otherwise.

Chapter 2 profiles the biggest discovery of its time, made by Cynthia Kenyon at the University of California, San Francisco, in 1993. This was followed by a second discovery in 1999 by outspoken MIT researcher Lenny Guarente, described in chapter 3.

Chapter 4 explains what happened when researchers learned they were looking at versions of human genes. Into the personal struggles of two researchers converged new business and science forces as their discoveries about ancient molecular genetic processes suddenly gained attention. With the attention came disputes, mistakes, and competition.

The idea met with objection but inspired disaffected thinkers and a few investors. Chapter 5 explores the first businesses of longevity in 2000–2003. Few discoveries moved so fast from such outlier status to the interest of finance while inspiring intense criticism.

Chapter 6 tells the story of the first compound to affect a human longevity enzyme, a trigger called resveratrol, discovered by David Sinclair, an outspoken young Australian graduate of Guarente's lab in his first years at Harvard. Sinclair started his own company, Sirtris, and raced the company of his former lab director to apply their discoveries.

Two genes became "celebrity genes," each gaining followers and critics who explored their many effects in hundreds of labs, described in chapter 7. Some significant proponents of the longevity idea were wrong, however, and provoked a powerful backlash in conferences, media, and university hallways. Part 1 concludes with the question: What happened when you took antisocial and brilliant scientists, threw a ton of money at them, and asked them to perform antiaging miracles?

1 *"Greater Than the Double Helix Itself": 1980–1990*

In all our exploration of longevity, going to the beginning of language and latest of human follies, lies a bargain. To overcome aging and death, most of us would give almost anything. That slinking wish lies at the foundation of much of our storytelling and many of our founding mythologies. The fear of death is so unbearable we traveled the world seeking a fountain of youth.

Partly for that reason, scientists have insisted that the processes of aging were random and uncontrollable. In 1825, the British actuary Benjamin Gompertz went so far as to quantify the depressing inevitability of our decline. He calculated that after puberty, the human death rate doubles every eight years. The older you get, the more likely you are to die. Some theory. There was no fountain of youth nor the possibility of a fountain of youth.

In the early years of the twentieth century, however, scientists applied the power of Darwin's theory of natural selection to the question of aging. They saw that the influence of natural selection fades over a lifetime, affecting only the hormones and behaviors that contribute best to an organism's chance to survive long enough to reproduce. Youth hormones like estrogen in women and testosterone in men are favored, because they grant reproductive fitness even though they may harm us later in life. By the middle of the twentieth century, this trade-off idea between youth and age had a name: "antagonistic pleiotropy." Pleiotropy means that one gene has many effects. Antagonistic pleiotropy means that virility or fertility genes that trigger youth hormones to keep us vigorous and attractive can later cause us to age more rapidly.

The discovery of such hormones led to a wave of early charlatans who transplanted goat and monkey testicles into their patients, preying on the eternal insecurities of potency in all of us. In the 1920s, the Viennese-born doctor Eugene Steinach promised that vasectomies would increase male longevity. Austria's leading scientist, Sigmund Freud, and America's biggest cynic, H. L. Mencken, both got themselves "Steinached." The Kansas-born John Brinkley, who twice ran for state governor, made a fortune by implanting goat testicles right beside the natural ones of his patients. Brinkley was so successful with his dangerous surgeries that he single-handedly gave life to the fledgling American Medical Association as it tried to stop him. When they did so, he took his radio personality to Mexico and founded the first of the great border-blasting broadcasters that gave the world rock and roll.

But what exactly causes aging? Several twentieth-century ideas sprang up to explain it. The genetic mutation accumulation theory suggested that it was unrepaired damage to DNA. Another idea, the error catastrophe theory, offered that cellular mistakes build until they reach a tipping point of disaster. Yet another theory, called hormesis, offered that a little stress improves longevity. The free radical theory of aging, which claimed that the reactive waste molecules of oxidation cause the body to break down when they bind to other compounds in the cell, became a widely accepted idea. But the main scientific point, driven home by serious research in order to counteract all the quackery, was that aging processes are always chaotic, disconnected, and uncontrollable.

Most researchers were influenced by evolutionary biologist George Williams, who in 1957 said that the processes of aging had to be random, "never due largely to the changes in a single system." His idea made the scientific quest for longevity unsavory. "This conclusion banishes the 'fountain of youth' to the limbo," Williams concluded, "of scientific impossibilities."

The discovery in 1965 by biologist Leonard Hayflick at Philadelphia's Wistar Institute that normal human cells in a cell culture divide only fifty-two times, never more, confirmed the inevitable limit to human life span. The discovery was even called the Hayflick limit. Scientists like Williams and Hayflick pounded out a jeremiad against the pop science of longevity. Their thought generated overwhelming doubt, which made studying the biology of aging an uphill battle for serious researchers.

There was one discovery, however, that tantalized the later generation of aging researchers. In the Depression, a Cornell University veterinary professor concerned about diet observed that when he trimmed back the feed of his animals, they actually lived longer than normal. In the era of soup lines and hunger, Clive McCay found that rats and mice lived 40 percent longer if you cut their feed by 30 percent. McCay published his longevity findings in 1934 in a respected journal and went on a long publicity tour. But he was not in the mainstream of aging research at the time. Some of the mice were sterile and many of rats showed reduced litters, so it was thought they had sacrificed reproduction for lengthened life. The assumption, quite reasonably, was that the long-lived, half-starved animals had a lower metabolism or a loss of fat.

A charismatic character who described himself as a "bit of a Bolshevik," McCay was a biochemist and professor of animal husbandry who planned the meal rations for America's World War II soldiers. He got himself a lot of publicity, but his caloric restriction idea languished. Few serious scientists wanted to pursue it.

By the 1960s, a contrary, ascetic mathematician and immunologist named Roy Walford latched onto caloric restriction in a best-selling book called *Maximum Life Span*. At UCLA, Walford became a low-calorie diet champion and practitioner who influenced millions, attracting followers like Timothy Leary and inspiring significant new researchers to enter the field. His work led to a cascade of scientific papers on caloric restriction in mice, rats, monkeys, and even humans, sparking the founding of the Calorie Restriction Society International in, appropriately, Las Vegas in 1994. But his best-selling books were often reviled by experts in gerontology. Walford passed away at the age of seventy-four, a gaunt figure permanently damaged by poor atmospheric conditions in his biosphere experiment in the Arizona desert. Still, he found that the biospherans' restricted diet had lowered their cholesterol, blood pressure, and blood glucose levels. He and his followers promoted the healthful effects of caloric restriction with science studies and popular books, and a following across the world came to include some important new biologists of aging.

Policy makers were thinking about aging, though, because they had to. By the end of the twentieth century, the percentage of Americans over the age of sixty-five was projected to grow from 20 percent to 41 percent of the total population. In 1974, Congress created the National Institute on Aging (NIA) as a new division of the National Institutes of Health in a nation suddenly aware that it was graying fast. How would the members of a developed country fare in a world that half a century earlier would have considered them lucky to make it to fifty-five? The American NIA pushed academic science to take on the big but unsavory biomedical quest to improve the quality of aging. The main concerns were the rising rates of cancer, heart disease, and dementia. The incidence of Alzheimer's disease, which got its name just ten years before, was expected to increase fourfold, up to sixteen million sufferers, by 2030.

In the late 1970s, a few academic conferences on the genetics of aging had sprung up. "If we found one thing, a trick say, that led to the mechanism by which longevity is achieved in mammalian species," said National Center for Toxicological Research biologist Ron Hart in the first wave of genetic idealism, "it would probably have a greater effect than the discovery of the double helix itself."

On a February night in 1977, a biologist named Tom Kirkwood was thinking about some of these issues, especially the trade-off theory of aging called antagonistic pleiotropy, while sitting in the bath in his northern England apartment. Kirkwood wondered how cells had made the same proteins unbelievably accurately for hundreds of millions of years. Cells in principle can be as accurate as they want, he reasoned, but at a cost of expending great chemical energy. Kirkwood loved thinking about big questions. In his bath, he asked himself a big question: How do we age?

As he sat in the steaming water, Kirkwood realized that the replication of seed cells like sperm and egg requires tremendous accuracy, but not so much the soma, or body cells. Sooner or later the body decays, so it does not need to be perfect. The most efficient way of assuring survival, then, was to devote super care to the seed cells and maintain the soma until the animal reproduces. The body was disposable. Eureka! He leaped out of the tub with the "disposable soma" theory of aging, *soma* being the Greek word for "body."

What Kirkwood hinted at was the difference between aging, commonly understood as the random breakdown of body tissues and organs over time, and life span or longevity, which could have some degree of genetic influence. Of longevity, even the noted scholar Leonard Hayflick admitted a few years later, "Evidence for the proposition that *longevity* is somehow determined by genetic events is overwhelming."

Such was the idealism of the first wave of researchers, utopian and generous but lacking much science, or money, to back it up. The problem was how to test any of these theories in an actual living being. Our many breakdowns, graying hair, weakening bones, and fading memory seemed too confusing to study scientifically, like the shifting eddies of a mountain stream. Were such breakdowns causes or effects? How would you ever separate the two? For that reason, the magnificent theories of aging or longevity amounted to little more than educated guesses more or less before 1980. To study them meaningfully required a short-lived, free-living, clear, beautiful, near-ubiquitous, voracious tiny animal. Enter the worm.

The "Gang of Cryptographers"

It all began with a tiny worm. The nematode lives virtually everywhere on earth, from mountaintops to deserts. In backyard mulch heaps and in the crevices of Antarctic mountains, in the stomachs of many large animals, nematodes are overwhelmingly the most numerous animals we know. They parasitize almost everything we eat, from sheep and steers to the cores of carrots and coffee beans. Four-fifths of all the visible life forms on earth are members of the nematode family, which counts among its branches an elegant, free-living curlicue named *Caenorhabditis elegans*, so elegant scientists made it part of its Latin name. Transparent, fluted, and graceful, it has a head and digestive and nervous systems, and yet is the size of the period at the end of this sentence. If you stare at one long enough through a microscope, you will see one of its 959 cells divide, which is, as one postdoctoral fellow once observed to me, "like seeing God." When the *Columbia* shuttle incinerated and crashed to earth, the only thing to survive were the *C. elegans* in silver-clad lab containers; they were found in a Texas field. To nab one with a pick under a microscope and move it from one plate to another, as I have done a few times, is a thrill of science.

In the 1970s at the Medical Research Council's Laboratory of Molecular Biology in Cambridge, England, a group led by Nobel Prize winner Sydney Brenner picked *C. elegans* as the model for a quest to study the way the nervous system affects behavior. Brenner, a short, fast-talking, thick-browed South African, had helped to discover messenger RNA and the code for the body's twenty amino acids. His dream of understanding the nervous system never panned out, but his choice of the worm as a lab model helped give birth to a new field in developmental biology. He inspired young researchers, recalled John White, professor of anatomy and molecular biology at the University of Wisconsin, "by talking nonstop about how he was going to transform science. Sydney never stopped smoking. My head was reeling." The molecular biologist Joshua Lederberg called them "the gang of cryptographers."

By the end of the 1970s, the group completed a remarkable timeline of every cell division in a worm's development from embryo to adult, work that brought the Nobel to Brenner and his colleagues Robert Horvitz, who went on to MIT, and John Sulston, who went to direct England's Human Genome Project. Horvitz discovered the genes involved in programmed cell death, or apoptosis (from the Greek for "falling away"), which offered a hint that genes timed the processes of life and death. Their timeline helped create a revolution in biological focus from "fixed entities" like genes, University of Illinois researcher Carl Woese wrote, to "fluid processes" like the translation of genes into proteins. Their gene analysis of the worm paved the way for the human genome race.

More than that, they shared an ethos. "We were an evangelical sect," Brenner said, "preaching to the heathen." New ideas and data were shared immediately in a free mimeographed journal called the *Worm Breeder's Gazette*, modeled on an English gardening magazine. One early cover featured a giant worm staring down through a microscope at a plate of tiny terrified scientists. At night they met at pubs like the Green Man in the town of Grantchester. On weekends they discussed politics at the Cambridge bookstore. They believed in research "as an unending argument between a great multitude of voices," the physicist Freeman Dyson later wrote, "in a continuing exploration of mysteries."

Their main discovery, already heralded by fly researchers, was of a small number of shared developmental genes that built similar body components in vastly different animals. No one expected that life followed such a unified, timed blueprint. The finding laid the groundwork for a new understanding of growth, Brenner said, as a "flow of information through a biological system."

The key to that flow of information is DNA, specifically, the protein switches that can work to unravel its tightly wound threads, process the information they contain, and translate that information into other proteins that do the real work of life. The longevity gene story is really a story of switches that give cells their work assignments. It is a switch that instructs a cell to become part of a nerve, or heart, or intestine. Without such gene switches, a body could not take shape from a two-celled embryo. Controlled by triggers that sense the environment, these gene switches instruct the cell as to which of tens of thousands of proteins to make and when to make them.

The process is hard to visualize and infinitely fascinating: Sticky and clear, easy for school children to extract with a cheek swab, DNA is easiest to visualize as a child's model. Two tubular strands run up the sides of a string of DNA with ladder rungs between them. When one strand separates from the other, a messenger makes a perfect copy of the original and rolls this template out to the ribosome of the cell. The ribosome in turn uses the template to make the proteins. The methods, triggers, and schedules of these tiny molecular switches unwinding DNA strands are the keys to the revolution in the biology of aging. DNA held the blueprint, but the instructors reading the blueprint proved to be the real keys to longevity. To understand that, an outsider needed to step in.

Gene Revolt

In Marinette, Wisconsin, north of Green Bay, Michael Klass grew up as the son of a telephone lineman. At the age of seventeen, he sat frozen through the famous Cowboy–Packer NFL Ice Bowl championship game. After studying engineering for a while at Michigan Technological University, he transferred to the University of Wisconsin–Madison in 1971, while the campus was in the midst of revolt against the war in Vietnam, where he wandered into the lab of early aging researcher Joan Smith Sonneborn.

One of the first researchers to use a simple lab animal to explore DNA's life-extending capacities, Sonneborn was studying aging in the paramecium. After sex, a one-celled paramecium near death will be reinvigorated into an entire second youth. Her research detonated an early Santa Barbara, California, aging conference, an observer noted, "like a neutron bomb."

Klass was fascinated. "Aging just seemed to me to be an illogical process, like a deleterious form of development," he said to me from his office at Abbott Labs in Chicago. "I wanted to know why life allowed it to happen." He went off for a postdoctoral position at the University of Colorado in Boulder to study longevity in the worm *C. elegans*. Working with the researcher David Hirsh, he focused on the worm's unique state of suspended animation called *dauer*, German for durable. At two days old or so, early in their development, juvenile worms face a turning point: when food is plentiful, they mature and have babies. But if food is lacking, signaled by powerful chemicals called pheromones, then growth stops and they enter the dauer state. In this suspended state, the worm can survive for up to two months. It can curl into a tiny ball, stick up its end to try to catch a passing bird, and wait to be transferred to a site with more food.

Hirsh won an early NIA grant as he and Klass discovered that worms live longer when you lowered the temperature. More important, when the animals came out of suspended animation, they lived a normal life span. The dauer state thus made a cosmic time-out from the aging process. The discovery landed them with a paper in *Nature*. But they did not get much response.

Klass wondered whether genes controlled life span. Gene mutations certainly limited life span. "If there was a mutation in a vital gene it could cause the death of the organism," he recalled, "but could you get mutations that would lengthen life span—and what would those genes look like?"

Klass started his search for long-lived mutants in the Hirsh lab in Boulder, working for four years across the hall from a friend, Tom Johnson, who also studied worm aging. By 1979, Klass moved on to a faculty position at University of Houston, where he won his own NIA grant to screen for mutants that lived longer, a very simple idea few had tried before. He identified five long-lived lab strains. Most were skinny and uncoordinated, even for worms, and he surmised their longevity was due to a defect in their digestion of food, giving them life extension due to caloric restriction. Given the crude technology of the time, he never sought to see if the life extension was being caused by a single gene out of the nineteen thousand spread on the worm's six chromosomes. It would be like taking a bad road atlas, adding a Great Lakes survey, and then trying to find a pipe in a bathroom in a house in Green Bay. "These results appear to indicate that specific life-span genes are extremely rare," he concluded in a report in the journal *Mechanisms of Ageing Development*.

By then, Klass's marriage was failing and he was thinking of leaving academia and returning to the Midwest to be nearer to family. He took a job with Abbott Laboratories and began a new quest for cancer diagnostics. "I really wanted to put my molecular biology knowledge to some practical use," he recalled. Although he would later go on to author some twenty-odd patents for cancer treatments at Abbott, the larger world largely forgot Klass's pioneering aging work. Klass called his old Colorado friend, Tom Johnson. Would he want some frozen long-lived worm mutants?

Midas's Gold

The round-faced, red-haired Tom Johnson was used to quixotic quests. Born in 1948, Johnson was raised by his mom and adoptive father, who was an ironworker and a difficult man. Johnson underwent the rigors of Jesuit prep school training in Denver. He later recalled of his childhood growing up in his grandmother's apartment building:

> I learned early on how to wait. My mom was 17 when I was born. She lived with her mom who lived with her mom. I grew up with a 65-year-old woman as the *maitre d'omo*. I was really interested in what a 65-year-old woman was all about. I consider myself to be very spiritual, that our lives have a purpose, and our job is to find that purpose.

He toyed with becoming a Catholic priest or taking an appointment at the Air Force Academy in Colorado Springs. Studying genetics in graduate school, he turned to his aging, beginning with flies at the University of Washington, and then with the nematode when he came to the University of Colorado. He had found his purpose.

Aging was an "intractable scientific problem," he recalled. "No other scientist would touch it with a ten-foot pole." Johnson won an assistant professorship at the University of California at Irvine, where he applied for a grant from the NIA, which was trying to get researchers interested in the basic biology of aging. In theory, genes extend lives all the time, whenever an organism masters its environment. Bats escape predators by flying; they live twenty times longer than their near-identical genetic cousins the mice. Brown squirrels adapt by climbing trees; they live twenty years to the three of their cousin, the rat. Elephants, tortoises, and whales live up to a century. But few were applying for NIA grants.

Tom Johnson wanted, like Klass, to use worms as a background to test the various hypotheses about aging. "The worm proved such a good model because of its clean genetics," he said. In his first year he taught a bright young undergraduate Californian, David Friedman, who was inspired to ask if he could work in his lab.

Johnson was publishing genetic analyses of aging, and he set Friedman to "figure out what was going on genetically" that was making Klass's worms live long. "It was a pursuit of truth," Friedman recalled. To their shock, every time they mated one strain of long-lived worms with normal worms, exactly one half of the offspring were long-lived. This result strongly suggested a single gene was responsible for their 50 percent–longer life. Friedman, who went on to a science career, recalled the reaction to his first undergraduate experiment:

> People looked at the result very skeptically. It was "this isn't real," or "there's something else going on." People said, that's why they're living longer, they're uncoordinated, they're eating less, so it's a dietary restriction effect. We were able to tease all that out by removing those genetic markers. We showed that it was not due to dietary restriction. It was not due to this, that, and the other. It was real.

To their mutual disbelief, they had found that a single gene was responsible for the longevity of some of Klass's long-lived worms.

Immediately, this obscure bit of science news hit a popular nerve. Tom Johnson was interviewed on the NBC *Nightly News* on November 17, 1986. Larry King, host of a talk show on the brand-new CNN, sent a limousine to his house. Johnson turned him down because he wanted science recognition, not popular fame. Johnson appeared in a *Los Angeles Times* profile in 1987 and in *Scientific American* two years later. They named the gene *age-1* and published the discovery in *Genetics* in 1988. After David Friedman left the lab, Johnson traced the mechanisms of the discovery and rushed his follow-up announcement, that *age-1* actually slowed the rate of aging, to the journal *Science*. "The molecular geneticists did not believe it. The evolutionary people did not believe it. I did not believe it," Johnson recalled to me, sipping coffee at a café on the island of Crete. "Still, probably because I was so skeptical, I felt I owed it to the world to get it out."

For an excruciating eighteen months, the journal *Science* held his article, but when it was finally published in 1990, Johnson hit the big time. The television newscaster Lesley Stahl told him his discovery was worth millions of dollars. "If you have really found a life-span gene," she said, "you have to invest in it." He spoke at the first West Coast Worm Meeting (yes, there are such meetings) at Lake Arrowhead, California, where Cynthia Kenyon heard him.

For all of the intense popular interest, Johnson did not get invited to speak at important scientific meetings. He felt shunned by the academic leaders in the field. "Every time I appeared on TV, my colleagues shied further away from me," he said. "People thought it was a crazy idea." Even though they had controlled for the animals' slight infertility, "I was never sure if they lived longer because they reproduced less," Johnson admitted.

Few biologists would believe you could have a genetic effect on aging or longevity, "let alone a single mutation," Johnson recalled. "That's the reason Mike Klass left academic science. I'm convinced he went into industry in large part because he couldn't convince anybody that anything we were doing had validity." To reassure himself, Johnson studied journalist Horace Judson' account of the original DNA explorers, *The Eighth Day of Creation: Makes of the Revolution in Biology*. "DNA is Midas's gold," commented one researcher of those old, fierce 1950s rivalries. "Everyone who touches it goes mad."

But now, important voices called for more research focus on the pure biology of aging. The University of Washington gerontologist George Martin explored genetic approaches to dementias. The University of Southern California's formidable Caleb Finch proposed that inflammatory responses that protect us in youth could cause the ravages of late life. These respected thinkers worked for more money for aging biology, but both railed against the emerging idea that tweaking one gene could replay life's oldest trick.

Johnson left Irvine to return to the University of Colorado for his wife's research. There he remembered his very earliest nights when he worked alone, back as a postdoc opposite Michael Klass's lab. Late at night a truck driver brought them discarded animal tissue from a slaughterhouse. The driver would see his light on and stop by to talk. One night, the driver asked about his research. When Johnson said he was studying a gene that extended worms' life spans, the driver became excited. From then on, he "would not stop bothering me," Johnson recalled. "Every time he made a delivery, he would ask me for my drug. When I told him it wasn't a drug but a mutation, he would say, 'Mutate me! Mutate me!'"

Maybe there was a fortune in it after all. The world was fascinated, primed even, but the idea needed a boost, a link to some respected theory, and someone with the ambition and courage to challenge what most everyone else said was true.

2 *The Grim Reaper: 1991–1993*

On the morning of her thirty-fifth birthday, Cynthia Kenyon awoke in a personal crisis. Her life was passing her by. It was February 1989, and she led one of the nation's top labs studying development in *C. elegans*. She loved her work in the towering University of California, San Francisco Health Sciences building. Her lab was small, but she supervised four graduate students on a search to understand the cell signals that helped all living things to grow and function. Kenyon had been tenured and promoted. She was a success. But she felt she haunted that she had not done what she wanted most.

When she heard Tom Johnson speak at Lake Arrowhead, it brought back a deep restlessness she had felt for years. Small changes in the activities of regulatory genes accounted for the differences among animals. In graduate school, she had learned that there were actually genes required for UV light to produce mutations. As with aging, everyone thought that UV just acted in a passive way, causing mutations by changing the DNA directly. But no—you needed a gene. Without this gene, UV still killed the cells, but it did not cause mutations. Something that everyone thought occurred passively was actively controlled by the genes. Could aging be like that too?

"I thought all the time about the Hayflick limit and the genes that time the growth of an animal," she said. She felt sure that genes controlled the rate of aging, and changes in these genes during evolution were "responsible for the different life spans of different animals." She became obsessed with the idea partly because it meant she would be doing something she had always wanted, exploring the unknown with the potential for understanding our own genes rather than those of tiny lab animals.

Everyone told her she was wrong. The biology of life span was subject to ridicule. Her colleague, biologist Patrick O'Farrell, discouraged her with theoretical arguments. When she suggested a conference on aging to another colleague, he told her there was no good work yet in the field. To change her own career to pursue its study "would be like sailing off the edge of the earth," said yet another colleague, now Harvard biologist and evolutionist Andrew Murray.

Kenyon was a well-regarded geneticist who worked on the genes that pattern the worm's body. Despite her success, however, she had a nagging feeling of being a one-hit wonder. In graduate school, she had exploded into the field with work that showed a new way to identify genes that were switched on by a change in the environment, called a promoter trap. After hearing a presentation, she came home talking about her idea for making these genes turn on like neon lights in response to chemicals that damage DNA. You could apply the same procedure to any bacterial cell to identify which genes are expressed or turned on. The promoter trap got her a paper in the *Proceedings of the National Academy of Sciences* and, eventually, her assistant professorship at UCSF. Her discovery was included in the book *Classic Experiments in Molecular Biology*.

But Kenyon worried that she had experienced the best of her science career by the age of twenty-four, and everything else would smack of nothing but disappointment. What she most wanted in her life was to study truth, and life span offered a way into truth.

Tall Trees

Cynthia Kenyon came from an old North American family. A Kenyon family Bible dating from the 1700s stood on a bookshelf in the living room. Her mother's mother was schoolteachers who grew up in the small town of Sharon, Connecticut. Her mother Jane's family, the Jacksons, were descended from William Bradford of the *Mayflower* and had a Connecticut street named after them. Her mother had been an equestrian, riding horses and competing against the William Buckley family "in their yellow barns and entourage," she recalled. Her mother, though, was a middle-class girl. When she traveled to compete, she slept in the corner of her horse's stall to save money.

Her father's family came from Ontario, Canada. Jim Kenyon's great-grandfather helped design the famous Lake Ontario Bridge, and his father, Pop, a doctor, was an award-winning oil painter and a horse-racing raconteur.

Cynthia Jane Kenyon started life on the South Side of Chicago in a row house opposite Stagg Field as her father completed his doctoral studies at the University of Chicago. She was four when the family moved to Teaneck, New Jersey, where Jim worked as an economist for the Port Authority of New York, flying a plane over Paterson, New Jersey, to chart every house, tree, and factory for his thesis on urban development.

Her mother was a medical technician who stayed home to raise the children, but with trips to the New Jersey Meadowlands to see the cattails and birds in the industrial swamps. Sundays her father took them to the Bronx Zoo and New York Museum of Natural History. Weekdays her mother drove them to Museum of the American Indian at Broadway and 155th Street. "Take your kids and go back to Jersey!" a taxi driver yelled when Charles, the youngest, ran in front of his car. In their row house on Fort Lee Road, up in the attic, one small window looked toward the George Washington Bridge to New York. Cynthia spent hours there with her sister Nancy, mesmerized by the glow of car lights. She watched the birds outside the window and read Kipling's *Just So Stories*, learning that nature followed logical and playful patterns if you just looked closely.

During the summers, the family escaped to her father's parents' farm in upstate New York called Tall Trees. They owned a huge white house surrounded by towering oaks, a painting of which she kept in her bedroom as an adult. All summer long the adults sat out back in the evenings and talked amid the gardens and birds, playing loudly competitive word games when relatives visited. In those nights Kenyon learned how to act in the world.

"How do you catch a wild bird?" she once asked her grandfather.

"Why, Cynthia," he said, "you put salt on its tail."

She ran into the kitchen to find a salt shaker. She waited by the bird house until a thrush flew in. Then she reached in quickly and returned, so the story went, with the thrush in her hand.

It was a story family members tell differently, but she was a child who had to go her own direction. "There's the right way," her father would say, "the wrong way, and Cynthia's way." When Jim Kenyon secured a geography professorship at the University of Georgia and the family moved to Athens, she became a "nature girl," said her friend and fellow musician Alicyn Warren. She would cross the Middle Oconee River and lie in a field of horses or wander in the woods behind the house, playing her guitar and writing songs in black music composition book. Kenyon took up the French horn seriously, making all-state orchestra and winning the coveted state scholarship three years in a row.

She wrote short stories and in high school published two in *Seventeen*. One was about a Vietnam soldier killing his first combatant. At the time, Athens, Georgia, was in an uproar in the battle for civil rights. She spent hours in the woods with her dog, sometimes camping out all night alone. She and her sister sewed their own clothes and strung chokers and beads. Kenyon kept a parakeet uncaged in her room, cutting down tree branches and laying out newspaper. She played solitaire at night and the bird would come down and take cards out of her hand and put them in her lap. On family vacations, "Her father would drive along, pointing out the reasons people lived on side of the river and not the other," said Alicyn, with whom she shared long discussions about philosophy or whether there is a god. "He drove her hard. The way she sees the world is largely due to him."

College turned out to be a bust at first. She was a professor's child who moved from subject to subject, seriously thinking she would be a concert French horn player, a writer, a French scholar, until she finally dropped out. Living in a farmhouse with a man during Athens's heyday of the rock-and-roll band R.E.M., she raised chickens and thought about becoming a dairy farmer or veterinarian. Cynthia Kenyon worked two jobs, waiting tables at a restaurant and ushering at the Beechwood Shopping Center movie theater, studying a French textbook with a flashlight while the movies played. Her mother watched from a distance, thinking about her daughter's love of nature and of complicated puzzles. She brought home for her daughter a textbook by James Watson called *Molecular Biology of the Gene*.

At night, Cynthia Kenyon thumbed through Watson's book, following the accounts of transcription factors, proteins, and amino acids, all the multiple players in a single cell. Reading chapters like "The Mendelian View of the World," "How Nucleic Acids Convey Genetic Information," "The Importance of High Energy Bonds," and "The Mutability and Repair of DNA," she came under the spell of the new field called molecular genetics. "Molecular genetics" described the incredibly complicated ways genes grant us the miracle of our eyes, hair, dispositions, and perhaps life spans. "I thought it was the key to virtually everything that was exciting in biology," she recalled. Reading the book was like peeking behind the curtain and finding the real, powerful players running our lives. Watson's was the field's first text and he wrote beautifully about genetic information being divvied up in cell division, the cellular processes of inheritance, and sex. Nature at the level of the molecule, she thought, offered a key to truth.

Kenyon reentered college at nineteen as a biology major and graduated as valedictorian. She won a fellowship to the doctoral program at MIT, where she worked in the lab of Graham Walker, a rising star who was studying the ways in which DNA-damaging agents can make bacteria more resilient. In his, lab Kenyon used a new "gene reporter" tool to learn that the DNA-damaging agents actually switched on genes, one of the first times anyone had looked for genes based on their activity. But she saw the coming wave of science to be developmental biology and switched fields completely to study the worm with Sydney Brenner in Cambridge as a postdoctoral fellow.

People were showing that just one minor change in regulatory genes could make a huge change in an animal. Change one gene and a fly will have an extra wing. The idea of such regulatory genes appealed to her philosophical side. Kenyon was fascinated by DNA's creative power and resolved to explore it in her lab at the University of California, San Francisco.

"A Perfect Vulva Every Time"

In California, Kenyon worked alongside future stars like Stanley Prusiner, future Nobel Prize–winning discoverer of prions, the agents of mad cow disease, and Elizabeth Blackburn, future Nobel Prize–winning uncoverer of telomeres, the chromosome endings that shorten as we age. Kenyon earned tenure for her work on a gene that built the segments of the body in the fly in the same order as the segments on the gene itself, discovering that this same gene also patterned the body of *C. elegans*. These findings demonstrated such genes, called HOX genes, were not simply involved in segmentation but instead were part of a much more fundamental patterning system. Writing up her experiments in prominent journals, she unleashed a brash style on the world with papers titled "A Perfect Vulva Every Time" and "If Birds Can Fly, Why Can't We?"

Kenyon taught her small lab group that science was art, as passionate and involving as music, painting, or writing. She took them to the San Francisco opera, riding her bicycle with her dress stuffed in the saddlebag. In the lobby, they took in the smells of perfume and glowing lights. Opera presented the human pageant in all its passion, dreams, unrequited love, and unlikely plot twists. She believed science could do the same. Friday afternoons they had a reading club like most labs, where they discussed new articles. But their topics ranged wider than most, including essays like that of the computer theorist Alan Turing on the engineering of biological systems. She studied business management techniques and gave her students responsibility for their agendas. They went on ski trips, performed skits making fun of her and the other professors, gathered on her deck and played music, smoked Cuban cigars, and convened for Saturday breakfasts of pesto eggs. "She sketched out my career for me on a napkin," said graduate student David Waring. "I felt I was learning things about life no one else knew."

Kenyon pushed to study aging. No one in her lab would do it. Her colleagues said there was no good work in the field. The university featured a program where students each did three three-month rotations in different labs to determine their final destination. For an ambitious graduate or postdoctoral student, the prospects in the search for longevity genes looked bleak. "The joke became if your experiment was not working, Cynthia would suggest looking for immortal worms," recalled graduate student Lisa Wrischnik.

Still, Kenyon thought that aging must be subject to gene regulation. "The genome is built in such a way that you can evolve rapidly by changing regulatory genes," she said. "Very similar animals, like the mouse and the bat, have very different life spans, so perhaps tweaking just a few genes can change life span." When she heard Johnson speak, "I had this deep emotional response," she said.

None of her advanced students would risk it. Aging was too complicated. It just happened. Their experiments would not work. There would be nothing solid to find. Finally, in 1992, a rotation student, that is, someone barely out of undergraduate biology, agreed to try. Ramon Tabtiang was a tall, sophisticated Thai American from an aristocratic family who was interested in argument and law, with a gambler's taste for the long shot. "This was the time of the great paradigm shift in biology," Kenyon later wrote for the Royal Society. "Organisms used highly similar molecular mechanisms, albeit with variation, to carry out the fundamental processes of life . . . . Ageing is a near ubiquitous phenomenon, and something so universal seemed to me likely to be regulated."

Aging and Hibernation

There were two ways to learn about the gene regulation of aging. The first was to identify changes in older cells, such as DNA damage, and ask whether any of these changes were not simply correlated with aging, but actually caused it. The other way was to seek lab animals that had altered life spans. Then one could observe them to see if aging in the mutants was regulated in a controlled or haphazard way.

In 1993, Tabtiang chose a temperature-sensitive mutant that had a defective gene controlling its entrance into the dauer, or alternative juvenile state, like a spore. Because normal worms give birth to some three hundred babies after four days or so, it would be impossible to find the parents amid all the offspring to see how long they lived. Tabtiang used the temperature-sensitive dauer mutant called *daf-2* (*daf* for dauer formation) and raised the temperature after they reached adulthood, causing their babies to go into suspended animation so he could easily chart how long the adults stayed alive. He planned to treat them at 20° Celsius with a chemical that caused a few random mutations, and then shift the mutants as adults to 25° so their babies hatched as dauers.

When he began with the control group, something funny happened. The control group, with the temperature-sensitive mutation, lived twice as long as normal worms. The adults moved well and reproduced, and looked healthy at an age when they should have been dead. He strolled into Kenyon's office. "You won't believe this, but the *daf-2* mutant is living long," he said. He thought he had worms with Tom Johnson's *age-1* gene lurking in the background. But when they checked they realized that was not true. Instead, he had discovered that the hibernation gene *daf-2* was a kind of Grim Reaper. Turn the gene down a lot, and the animals never grow up. Turn it down a little, with a weak mutation, and the animals lived 100 percent longer than normal—the longest life extension ever seen in a lab.

It was only a worm. But to Kenyon it was exactly what she had been thinking all along: there were regulatory genes for aging. A life-extension program could be turned on in a normal adult animal. The discovery opened a "big door," said her colleague Ira Herskowitz. Because the long-lived mutants also had 20 percent fewer offspring than normal, she had to test if the longevity was a result of reduced reproduction. Another rotation student, Jeanne Chang, learned to sterilize the worms and found that sterile *daf-2* mutants lived much longer than sterile control worms, meaning that reproduction had nothing to do with the longer life span.

More important, the long-lived animals required a second gene called *daf-16*, which had been discovered by the original research into dauer activators, and which superintended the processes lengthening life. Kenyon took the plates into her private office, with its *Alice in Wonderland* and Edgar Allan Poe volumes on the shelf, to figure out what exactly the gene did. They nicknamed the gene "Sweet Sixteen" because it did the actual life-extending work both in their worms and, they later learned, in Tom Johnson's *age-1* mutants. The gene "extends the life span of . . . adults by triggering expression of a regulated life span mechanism," Kenyon wrote with gusto in a letter for *Nature*. "The magnitude of the life extension was striking," she wrote on December 2, 1993, "the largest ever in any organism." They had discovered a full-blown longevity "program that can, if activated, prolong life substantially." The discovery provided "entry points into understanding how life span can be extended."

Nature commissioned two experts to comment, and they were skeptical. "Everything we know about the evolution of aging suggests that it is probably the most polygenic of all traits," wrote University of Edinburgh evolutionary biologist Linda Partridge and Oxford zoologist Paul H. Harvey, meaning the process was affected by many different genes as well as chance events. "A regulated mechanism of aging," they concluded, "is not expected to exist."

"If a Worm Can Do It . . ."

Kenyon felt she had discovered something unprecedented: that a single gene modification could change life span by changing the tools that regulated it. "It was a whole new concept," she said. "I felt we were on the edge of something. You could now look for longevity control genes." Her reaction was visceral. "You have on the one hand a normal worm, which looks awful at age two weeks, and the mutant worm, which looks wonderful. You sit and think, you really have gone where no one would have. I felt like I wasn't allowed, but I went anyway. I saw this thing I wasn't supposed to see. That you could stay young, that it was possible. I thought, oh gosh, what have I done? The next thought was, I want to be that worm. If a worm can do it, I could do it."

Most molecular biologists and gerontologists believed she was taking the discovery too far. "We thought it was kind of a dumb result, that you were basically turning a healthy worm into a sort of partial dauer," recalled Harvard geneticist Gary Ruvkun. "Most of us said it's not that interesting." The idea that a single gene could affect the life span of a mammal was "incredibly naive," said the University of Washington gerontologist George Martin.

To the larger world, however, Kenyon's short paper detonated like an explosive. Researchers may not have agreed with her, but they had to pay attention. "People go all their scientific lives without making a discovery, and here she made one on her first try," recalled her former student Andy Dillin, later a Paul Glenn Center for Aging director at the Salk Institute. "It was a mixture of the workmanlike and the beautiful," said University of London biologist David Gems. "It remains the single most important aging paper in science." For her passion and enthusiastic writing style, different people outside the field, in business, onlookers and fans, resonated to its implication. "She was right," said Ruvkun, a little reluctantly. "It was interesting."

Then came another clue from an even more obscure organism.

3 *Sorcerer's Apprentices: 1989–2000*

In 1989, twenty-two-year-old Brian Kennedy was driving home from Chicago to Kentucky, after graduating from Northwestern University and three weeks after getting married, all set to go for a doctorate in biology at MIT. He and his new bride were exhausted. Kennedy, an ex–high-school golfer from Louisville who loved reading Russian novels, was thinking about the next step in his life. In the dark descent across north-central Indiana, a blacked-out car was heading straight toward him. In the next instant, Kennedy was crushed against the steering wheel. They had smashed head-on with a drunk driver.

His wife was fine, but Kennedy spent the next six months in a wheelchair. "I lost a year and almost died," he recalled. "It convinced me that you never know what's going to happen in your life. I decided to go for it, not be timid, and give science my all."

After three surgeries, Kennedy was anxious to get going with his research. At MIT in his first semester, he was taking classes when he met a graduate student in the beat-up student lounge nicknamed "The Pit." Nic Austriaco was a Filipino with a spiritual bent. In the building called "E-56," they hung out on the old couches and discovered they shared an interest in big philosophical questions. Austriaco had an encyclopedic mind, and Kennedy had experience working with yeast. In January 1991, they walked into the office of a professor named Lenny Guarente to see if they could pursue something big.

At MIT, Lenny Guarente (no one ever called him Len or Leonard) was a forty-five-year-old, soft-spoken, tall, bearded leading expert in the process of reading and copying DNA to make proteins. Guarente studied the process in yeast. Like *C. elegans*, baker's yeast was a well-known science research model, a fungus used in cooking and wine making since prehistory. Easy to culture in the lab, its cells are surprisingly similar to our own. Yeast produces the carbon dioxide that makes bread rise and wine and beer ferment and made almost as much a fundament of civilization as fire.

"Okay," Guarente said. "So what do you want to do?"

A long silence followed. They had not really thought out what they wanted to. The slight, dark-eyed Austriaco mentioned apoptosis. From there, they moved on to discuss the biology of aging.

Guarente had never even considered aging a scientific problem. "It seemed like something that just happened . . . like erosion," he later recalled. But he liked Kennedy and Austriaco and needed good doctoral students in his lab. Guarente was himself facing a turning point. His marriage was failing. He was scrambling to raise his three-year-old son on weekends. Guarente was fed up with the slow pace of his study of yeast transcription. He hosted an annual Christmas party for his graduate students at his house and found in his students the family life he coveted. He later covered his lab door with photographs of those parties.

Once, early in his career, Guarente missed a big idea that he regretted never pursuing. It was cancer, a field that had ignited during the years he had toiled at the humdrum world of yeast transcription. He was willing to entertain something "out there," said Kennedy, like aging biology. He told Austriaco and Kennedy to read up and come back to him.

In their weeks of reading, Austriaco and Kennedy found papers written by Louisiana State University's Michael Jazwinski, who inserted mammalian proteins into yeast to see if they changed life span. They studied the Klass worm papers and the paper by Michael Rose, who bred long-lived fruit flies. Rose did so by "tricking evolution," forcing the flies to wait until later to reproduce, thus selecting for offspring that lived longer. Rose proved that longer, healthy life in flies could be inherited, meaning that it was partly coded in genes. "Aging is in no sense any basic feature of cell biochemistry," the outspoken scientist announced in a crusade for the study of longevity genes. Rose berated naysayers in a later *Discover* magazine interview, citing Jimi Hendrix dying at twenty-seven and Mozart at thirty-five as inspirations:

> There are all kinds of people who are opposed to us doing anything [about aging]. . . . I have heard on many occasions people give very moving addresses as to why we should die as soon as possible. . . . I just know other people who don't want to die, and least of all by the hor-

rible and unattractive process of aging, and I don't see any reason why they shouldn't be allowed to go on living.

Kennedy and Austriaco had stumbled on to something that incited passions. It was not that they were worried about longevity. Rather, "it seemed like aging in yeast could give us a new paradigm," Kennedy said. He added drily, "It exceeded my expectations."

They decided to take the opposite approach from Jazwinski's. Rather than insert mammalian proteins into yeast, they would study yeast aging genes and see if they existed in mammals. They hardly dared to hope to find a single gene mutation like Johnson's *age-1*. But yeast offered certain advantages: it was so simple and short-lived they could answer questions about long life genes quickly. "Intoxicated by this thought," Guarente wrote in his autobiography, *Immortal Quest*, they decided to pursue the idea.

Maybe aging was on the verge, as cancer research had been at the time Guarente was in graduate school. All of them were competitive and came at subjects from an outsider's perspective. Austriaco, for instance, raised in Thailand, approached problems in a unique way. His professors used to give him exam problems before other students to make sure there were no shortcut solutions that they had not anticipated. Guarente gave them a year.

On the Fairway

It was risky to think that one-celled organisms might age in a process applicable to mammals, but apoptosis had followed a similar trajectory and proved to be a fascinating topic in cancer study. Cells are programmed to die as part of the normal progression of life. It was a little like the old men who sat around the stoops in Guarente's hometown of Revere, Massachusetts, watching the street, sipping coffees.

Guarente's childhood made him a good mentor for students with a risky dream. Born June 6, 1952, Guarente grew up in a big extended family in a working-class town. Revere featured the country's first public beach, which had deteriorated by the 1960s into honky-tonk bars with a greyhound racetrack. Guarente's father was an underachieving Westinghouse clerk, his mother a schoolteacher. His older brother left home after an affair with a married woman. When Guarente scored well on a junior high science aptitude test, his mother pulled him out of public school and sent him to a Jesuit prep school. Traveling an hour each way by bus and train, he studied Latin and calculus and, for the first time, began to think he could be someone. He graduated as valedictorian.

Guarente made it to MIT as an undergraduate in the fall of 1970, where he took the seminar offered by future Nobel Prize–winning biologist David Baltimore. Inspired, Guarente went on to earn a PhD in molecular genetics from Harvard. At the end of graduate school, he joined Mark Ptashne's lab in the old Harvard Bio Labs Building on Divinity Avenue. The entranceway featured two giant rhinoceros statues the students sometimes dressed up. Ptashne was an expert in the new DNA technology that allowed researchers to recombine the building blocks of genes. Guarente's postdoctoral work there made an amazing introduction to the genetics revolution.

The quest to understand the genes of yeast aging was hard work. The first thing Austriaco and Kennedy had to figure out was how to measure life span. There were two ways to do so—either by counting the number of times a mother bud reproduces—about twenty-one—or by witnessing its actual death, which was hard since a mother is surrounded by hundreds of offspring. They decided on the former. Yeast gives rise to daughters by budding, and the daughters needed to be picked away every ninety minutes. Working twelve-hour, day-and-night shifts seven days a week through the summer in a hot lab that smelled like a brewery, Austriaco and Kennedy exhausted themselves. Eyes bloodshot, his hair flying in all directions, Austriaco could barely keep up. Finally, they tried refrigerating the yeast at night and studying them again in the morning, allowing the scientists to sleep. "It saved my life," Austriaco recalled.

Within a few months, they had found one particularly long-lived mutant strain. The question was, how could one determine if a single gene was responsible?

Guarente and Kennedy liked to golf the famous Brookline, Massachusetts, public course, where the very first U.S. Open was played in 1910. They discussed the problem as they walked the municipal course fairways. Somewhere on a wooded path, amid the rhythmic soft clack of the ball and the smells of grass and a sea breeze, they hit on the same idea: their long-lived mutant was sterile. If they could identify the DNA fragment that restored fertility, they might have the same fragment that gave it its longevity.

When Kenyon published her *Nature* paper, it shocked them out of their golf afternoons. Longevity genes existed, but somebody else was way ahead of them. They isolated the DNA fragment causing infertility and dubbed it Youth (UTH) with four variations. Austriaco studied Youth 1 and 3. He read the Bible and considered a different career. His genes proved a dead end.

Kennedy studied Youth 2 and 4 and hit the jackpot with Youth 2, which turned out to be a known gene called *SIR*, silent information regulator, so named because it had the effect of silencing other genes. *SIR* formed a trio of related genes that created yeast hermaphrodites—fungi with both male and female characteristics—and thus the mutants were sterile. It was a gene silencer.

When it turned out that the forty-second mutation of Kennedy's assigned gene made yeast cells live 30 percent longer than normal, they rushed to publish their result in the premier journal *Cell* in 1995. A mutation in a single gene could delay aging in yeast. It was the world's third bona fide longevity gene. It took them four years rather than one, and it was only yeast. Guarente's Nobel Prize–winning department chair Philip Sharp was not encouraging. "You were doing great things," he told him. "Why go off into this field where nothing's going on?"

A "Molecular Cause of Aging"

Despite the response, Guarente resolved to go after aging full bore. They knew the Sir protein moved from one location to another in the cell to lengthen its life; they just had to determine exactly where. Kennedy graduated and decided to go to Switzerland where a lab had the technique for lighting up proteins in yeast so one could detect their location.

While he was in Switzerland, Kennedy stained the Sir proteins with rabbit antibodies, the immune system workhorses that helped lab researchers identify molecular targets, and then with fluorescent goat antibodies to attack the rabbit, so he could see where the gene product was acting in their long-lived mutant. It turned out to be the nucleolus, the smaller organelle inside the cell nucleus, where the protein preserved genome stability by protecting ribosomal DNA. Ribosomes are the place where proteins are made, and they contained a small vestige of DNA from billions of years ago when the tiny cell structures were once free-standing organisms.

Back at MIT, Guarente's lab figured out that eventually the damaged ribosomal DNA snippets broke off and formed rings, adhering one end to the other. The rings seemed related to yeast aging. Guarente was getting more excited, but he had no help. Kennedy lingered in Switzerland, and Austriaco was about to leave science to become a Dominican priest. Guarente postdoc Brad Johnson was looking at the rings. He proposed an experiment to verify a theory that ribosomal DNA loss drove aging.

Around that time, Guarente traveled to Australia to talk about his discoveries and ran into a David Sinclair, a focused, brash young doctoral candidate at a dinner. Sinclair was a short, sandy-haired student working on his PhD in molecular genetics at the University of South Wales. He liked to break rules. He amassed so many speeding tickets in his red Mazda Miata that it was confiscated. When he heard Guarente, he thought, "You did not have to understand everything about aging, you just needed one gene intervention. I said to Lenny, 'I'll take out a loan, I'll sell my car' to come to the lab." In 1995, with a Thompson Prize for best thesis in hand, Sinclair applied to come to MIT, and in 1996 Guarente accepted him.

The lab atmosphere intensified. "There were ideas and debates flying through the air constantly," Johnson later said to *Science* magazine. Amid the debates, some of the men in the lab posted sexually suggestive notes as a joke. In the final escalation, Sinclair pinned on a door a poster of kangaroos mating while a shocked couple watches, the last straw that led to a policy change. He annoyed people but showed up at 8 a.m. and often stayed past midnight, catching the last "T" train home. He headed out for beers on Friday afternoons, but on Saturday and Sunday he would be back working in the lab.

The life-span mystery grabbed him. At first, he thought that the aging process got rid of the ribosomal DNA circles as the early theory suggested. One day, Sinclair burst into Guarente's office. He proposed the opposite: during the aging process, cells accumulate ribosomal DNA circles rather than purge them, causing a toxic effect.

The idea was so significant he mailed it to himself in a sealed envelope to date his claim of credit. With technical advice from Johnson, Sinclair isolated aged yeast cells out of the tens of thousands of their offspring and analyzed their ribosomal DNA circles. He discovered that the aging cells indeed accumulated rDNA circles. If he induced more rings to appear by a genetic trick, the cells aged even faster. They published the discovery in *Cell* in 1997. For the first time, they wrote, "a precise molecular cause of aging . . . had been found." (When the *Cell* article appeared, lab members were surprised to see Brad Johnson's name left off.) The article triggered a mention in the *New York Times*.

At that same moment, the California biotech company Geron discovered a clue to the genetic mechanism of aging in human cells. Since the early 1990s, several companies had tried to influence telomeres, the chromosome endings like shoelace ends that shorten each time a cell divided. In Menlo Park, Geron said it had figured out a way to inject more of the enzyme that repaired the gene endings, called telomerase, into human cells. It seized the momentum of the newly emerging field as major scientists like the University of Denver's Tom Cech and the Whitehead Institute's Robert Weinberg, raced Geron researchers to clone or identify the human telomerase gene.

With the spotlight on telomeres and human interventions, Guarente's lab struggled to figure out the mechanisms for their lowly yeast gene. In 1999, two newcomers, one a Washington State postal worker's son who loved to challenge authority and the other a born teacher, pinpointed the gene *SIR2* as the key player. A Johns Hopkins Medical School researcher found analogs of the gene in mice and humans, a promising sign. "Genes bearing sequences very similar to *SIR2* were present everywhere," Guarente observed. It appeared in key reactions in everything from bacteria to humans. Spanning a billion years of evolution, it had to be of critical importance, Guarente felt. Most in his lab disagreed, but suddenly other researchers were trying to decipher the gene's method. What exactly turned it on, and why, and how? They found themselves in a full-out competition to find out.

The Sensor

In 1997, a dapper, intense, and friendly Japanese postdoctoral researcher, Shin Imai, arrived in Guarente's MIT lab. He was a funny presenter given to karaoke with his wife, but he was obsessively precise in the lab. Once when a fire drill threatened to stop a key procedure, Imai carried out his ice bucket, test tubes, and notebook and conducted his experiment on the sidewalk of Galileo Way as the cars roared past.

With another Harvard lab hot on the trail, Imai sought to figure out the *SIR2* mechanism and trigger. "We were like medieval time travelers," Guarente later wrote in his autobiography, "who saw that gas propelled cars but with no idea how." To make a precise measurement of the mass of the histone tail, or protein switch, they thought the Sir enzyme targeted. Imai and Guarente used mass spectrometry, a technique for measuring what's in a molecule by determining the mass and charge of its fragments. At the MIT Cancer Center's protein core, like "something out of NASA with its blinking lights and high tech equipment," Imai inserted the protein tail into the mass spectrometer. The machine sent the protein through a charged field to a sensor like "booting a football between two goalposts," according to Guarente.

Genes are wrapped around proteins called histones. The reason that a miraculous ten feet of DNA can pack so tightly into every one of the trillions of cells of your body is that the histone works almost "like the spool of a tightly wound gardening hose." A gene is activated, or turned on, when a chemical called an acetyl makes the gene unwind so that messenger RNA can be created. Remove the acetyl and the DNA condenses, becoming inaccessible.

The mass spectrometer gave Imai a puzzling answer. Their ending histone tail, or control switch, weighed less than the starting one. This was surprising since they assumed the protein was adding a small chemical. The missing weight was exactly that of a single acetyl group (42 daltons). Imai thought they had done something terribly wrong. The answer was too perfect. They sat, frozen.

Then Guarente laughed. The result was indeed precise. Their protein was *removing* an acetyl group from the histone. It was a histone deacetylase, or gene silencer.

More important, Imai discovered that a common coenzyme called NAD (nicotinamide adenine dinucleotide), an essential chemical synthesized from vitamin B_3, found in all living cells and critical to many metabolic reactions, triggered the silencing. In well-fed cells, NAD is unavailable. In starving cells, the production of NAD skyrockets, stimulating *SIR2* to silence genes. "Such an enzyme had never been described before," Imai later said, because it linked metabolism to aging by sensing nutrients. They thought they had found the key to explaining why reducing calories extended life. Imai wrote it up in his diary: "October 18, 1999. Sir2 is a NAD-dependent deacetylase!"

They rushed their paper to *Nature,* announcing they had discovered "a biochemical mechanism to set the pace of aging to the availability of nutrients." The paper's claim was so unusual that a reviewer slammed the report. Guarente begged the editor to overlook the lone dissenting reviewer, and the article appeared in 2000.

Media interest again perked up. Guarente was profiled in the *New York Times*. "Dr. Guarente has seized at one end of a highly interesting thread and is tugging with gusto," the writer Nicholas Wade observed. "The discovery could have very broad consequences because it identifies a cellular pathway—and possible drug target—by which caloric restriction works." The *Times* mention led to an appearance on *Good Morning America*, on the same show as Guarente's favorite blues singer, Eartha Kitt.

But not everyone was impressed. "I think Guarente's whole set of prejudices is wrong," commented fly biologist Michael Rose in the same article, insisting along with many others that there could be no single mechanism controlling the rate of aging. Still, Rose added, "it is relatively harmless if he wants to bash away at it."

"To Overthrow the Present Intellectual Order"

The discoveries suggested something counterintuitive. *SIR* silenced other genes. Keeping genes silenced is an essential activity in the body, since "chaos would ensue if all the genes in a genome were switched on in all cells," as *New York Times* science writer Nicholas Wade later described Guarente's work. Gene silencing makes a kidney cell know it is a kidney cell, a muscle cell know it is a muscle cell. Aging could be sped up or exacerbated by genes being *too* active. "Evolution overengineers for survival," Berkeley biologist Judy Campisi noted of active genes. "Underengineering might extend life."

In a similar way, Kenyon's Grim Reaper gene had to be turned down, triggering a "hunker-down state," she said, as dauer formation was triggered by a pheromone signal that nutrients were lacking. In 1996, *Cell* commissioned Kenyon to write a review article uniting the discoveries. She could not resist her practical joker's side, titling the article "Ponce d'elegans" and including her PowerPoint slide comparing the life spans of a mouse, a canary, and a bat. They were all upside down because they were dead. Because bats hang upside down when alive, she drew him right side up. The *Cell* editors were upset when they realized what they had printed. "The process of aging influences our poetry, our art, our lifestyle, and our happiness," she began, "yet we know surprisingly little about it." An organism's life span could be "raised to a significant degree," Kenyon wrote, offering the prospect of a similar longevity pathway in humans. "Perhaps the strongest argument [for that]," she wrote, "comes from the remarkable level of evolutionary conservation that has been found to exist in other biological processes." The signaling pathways calibrated the rate of aging all the way from birth to death.

In the same issue, Berkeley biologist Judith Campisi wrote another review discussing the immortality of cancer cells in humans. Campisi wondered whether cancer cells had been taken over by the inflammatory response gone haywire, essentially, and that the body defenses which protected us in youth drove many of the diseases that afflict us in old age. For the first time, new lab techniques allowed science actually to test aging theories in living animals and to separate cause from effect.

The yeast and worm discoveries "completely transformed" the new molecular study of aging, Kenyon and Guarente wrote later in a special *Nature* review issue titled, "Aging Research: The End of the Beginning." The biology of longevity was undergoing a "paradigm shift," they suggested, inferring that the rate of human aging could be modified by tweaking genes and understanding discoveries in small animals. Single gene mutations lengthened the span of healthful aging significantly, up to 100 percent in worms and 50 percent in yeast. If you added the studies of telomeres, oxidative damage, and animals in the wild, the new "scientific picture made an intriguing jigsaw puzzle," concluded the journal editors. No one knew if it would fit together.

The search for longevity genes took off in multiple laboratories. In the worm, four researchers turned to the newest tools of molecular genetics, gene mapping, cloning, and analysis to uncover exactly where and what the hibernation gene did in real time to double healthy life span. In yeast and the fly, others followed.

As the stakes increased, so did the competition. That was before the researchers knew what they had.

4 *Race for a Master Switch: 1995–2000*

Around the world, the Kenyon and Johnson discoveries caused a tremor in the minds of disaffected thinkers. The skeptical *Nature* reviewer Linda Partridge said at the Royal Society's 350th anniversary, seventeen years after Kenyon's paper, "It was the crucial mutagenesis. It knocked my socks off." The news spread to a former Gulf of Mexico oil rig supervisor, Cindy Bayley, a pianist with an MBA and PhD in biology. In Rio de Janeiro, chemistry student and track star Tomas Prolla was so enthralled by it and Walford's book that he wrote to the University of Wisconsin–Madison's Rick Weindruch to study aging in calorie-restricted rhesus monkeys at his lab. In Leiden, Holland, the researcher Eline Slagboom thought, here was vindication for her early work in humans. In London, the postdoc David Gems read Kenyon's experiment and changed his career path. These papers inspired graduate students to consider a career in the biology of longevity.

It was not simply the fact that longevity could be extended, it was also the particular way Kenyon said it. "Single gene mutations extended life span significantly," she declared in conference presentations. "The quality of life of the long-lived mutants is high. If growth or pattern formation is regulated by the same genes in all animals, then the genes that extend life in worms should do so in humans as well." A follow-up 1996 *Cell* review challenged the field: "We have no idea how many life-span genes will be found to exist," she wrote. "We should get going before it is too late."

She offended some people. "There's something manic about her optimism," said one of her lab's graduate students, Jeanne Harris. "People look at this woman with long blond hair, who doesn't talk or look like a scientist, and don't know what to make of her." Her proselytizing annoyed some of the researchers studying the hibernation and growth genes in the tiny worm, even those who were her friends, among them, the University of Washington's James Thomas, the University of British Columbia's Don Riddle, and Harvard's Gary Ruvkun. "I thought, Oh gosh, now I'm in aging research. Your IQ halves every year you're in it," Ruvkun told a *Wall Street Journal* reporter. No one knew the life-extending functions of *daf-2*, the Grim Reaper, or *daf-16*, the life-extending Sweet Sixteen. No one knew their locations on the worm's six chromosomes. People knew virtually nothing about their biochemistry. All they knew was they were dauer genes that caused a prepubescent animal to go into suspended animation when food was lacking. To figure out the genes' functions one would have to find them, clone them, sequence them, and compare the sequences to those of known genes.

The trouble was that with early 1990s technology, each of those tasks could still take months or years to accomplish. An all-out effort to sequence, or list, the complete catalog of worm genes was meant to show the way for the accelerating corporate and government race to solve the human genome. But the worm sequencing effort moved slowly. A science revolution, wrote the philosopher Thomas Kuhn, changes the "directions essential for mapmaking" created by "extended exploration" of an anomaly. Like the maps of the early European explorers, the worm gene map featured dark zones. The newly discovered long-life gene and the life-shortening gene lay in the dark zones.

Leading the race to clone the two genes was Gary Ruvkun, a laconic, Berkeley-educated molecular biologist and associate professor at Harvard. As a child, Ruvkun loved reading the classic 1960s *Time-Life* children's book called *The World of Science*. His father was an engineer who helped build the Oakland Coliseum. Growing up Jewish in Oakland, Ruvkun absorbed the grandeur of NASA's televised launches and the outsider "warped perspective" of late-night-television Jewish comics. After majoring in physics at Berkeley, the tall, dark-haired, skeptical thinker traveled the Pacific Northwest, where he worked as a disc jockey and planted trees, some that grew ninety feet tall. With a friend, he hitchhiked around Central and South America, floating down the Amazon before applying to Harvard as a graduate student in molecular biology. There he studied bacterial and fungal infections with the molecular biologist Fred Ausubel, eschewing some of the other Harvard big names Ruvkun considered self-aggrandizing.

Around his lab at Massachusetts General Hospital, Ruvkun, with his huge crop of black hair and horn-rimmed eyeglasses, often wore jeans or work pants and a motorcycle repair T-shirt. He talked with students for hours about sports, readings, ideas, their favorite obsession—anything beyond their work. He called the talks "drifting" and considered them essential to science. His students nicknamed his dark, book-strewn office with the vintage supermarket edition of *The World of Science* the "Black Hole," because once you went in you might never get out. He gave students Boston Red Sox seats from his season tickets. "He really cared about us," said Canadian graduate student Heidi Tissenbaum. "Gary taught us to be fearless," said graduate student Jason Morris.

He had many interests. Ruvkun yearned to join NASA's quest for extraterrestrial life. For years he had been studying dauer formation and the genes that timed the worm's growth. He would later discover the role of micro-RNAs in the animal's development, bits of genetic material considered a clue to the origin of life. But by 1995, after Cynthia Kenyon showed that a weak mutation in a hibernation gene he was studying doubled life span, he did what any good scientist would do. He jumped on it.

An intense personal rivalry drove the two researchers. Kenyon's discovery thrust Ruvkun's field into the headlines. He suspected Kenyon's longevity claim, because he felt she was crossing a line between science and "snake oil. I just try not to push it so far," he said. But the National Institute on Aging, with its "long shot bets based on two or three minor papers," got him going. "Until then, fruit fly genetics had been the hottest field in biology," he told an interviewer. They were trying to emulate that "luminous fly work on how development is controlled." Now, he admitted, "I'm into aging."

Insulin

For years a talented graduate student in the Ruvkun lab sought to analyze *daf-2*, the Grim Reaper gene, for her PhD thesis. Heidi Tissenbaum had a masters in physiology from the University of Ottawa, but had never used some of the basic gene-identifying technology before. "All these people were so far ahead of me," she recalled of her first years at Harvard Medical School. "I kind of squeezed by." Careful and committed, she worked nonstop on the gene, not leaving the lab for months on end. When she read that Kenyon found her gene was a life-span gene, she panicked. "I worried she was going to try to clone my gene, but she didn't," Tissenbaum recalled.

She had tried to clone the gene onto a man-made chromosome that could be used to create copies of large chunks of DNA. That did not work. "The technology wasn't good," she said. "I was bringing all sorts of new techniques into the lab." Tissenbaum spent three years trying to bind known DNA snippets to one end of the gene. She tried approaching the gene by "walking in from the chromosome ends, eliminating more of the ends bit by bit," she said. That approach failed. In her annual reviews, Ruvkun defended her. She began having dizzy spells.

To find the Grim Reaper gene, Tissenbaum had to run the worm's DNA through a gel on an electrified mixing machine in a process called gel electrophoresis. Gel electrophoresis works because DNA has a negative charge. The DNA oozes through a gel from one side to the other, separating the molecules by size based on the distance they move. Once they found it, they had to reproduce the gene. Only then could she determine its sequence of parts and compare it to known genes on a computer.

It was important that she succeed. It was known that the *age-1* mutant needed *daf-16* to live long. In 1996, the University of Washington's Jim Thomas discovered that Johnson's longevity gene, *age-1*, was actually a known dauer gene called *daf-23*, a discovery confirmed by Ruvkun the following year. Now it was clear how the Kenyon and Johnson discoveries were related. The worm's ancient signaling pathway did more than arrest development or cause hibernation. It extended life.

Then Tissenbaum realized why she was dizzy: she was pregnant. While she went on maternity leave, postdoc Koutaru Koumaru took over the *daf-2* search. He took the DNA and performed a kind of molecular cat's cradle, wrapping a string tighter and tighter around the region until, they hoped, it would be around the gene itself. Working late nights, referring to the pieces of the worm gene sequence being published in England, he found the gene right where Tissenbaum had mapped it, on the left arm of chromosome 3. They analyzed the gene sequence or list of its chemical parts and ran it through a computer search comparing it to all known gene sequences in a data bank. "You read these as they are ratcheting by," Ruvkun described the lengthy lists of chemicals to the *New York Times*. "MTV is good training." They were amazed to discover that the worm's gene resembled the human gene for the insulin and insulin-like growth factor receptors. The Grim Reaper was a human gene. No one even knew the worm produced insulin. "That changed everything," observed Tom Johnson. "I went from total skeptic to suspecting a human intervention was possible."

On August 15, 1997, the discovery dazzled the world with the "tantalizing possibility," wrote a reviewer in *Science*, "that changes in glucose metabolism could be the key to slowing the aging process in . . . humans." Long-lived mice and rats under caloric restriction had lowered insulin levels and higher insulin sensitivity. If applicable to humans, "*that* would be a phenomenal discovery," commented the gene's original discoverer, Don Riddle. Ruvkun appeared on the NBC *Nightly News*. "The excitement surrounding the idea of the genetic regulation of life span," wrote later lab postdoc Sean Curran, "evolved into hysteria."

Insulin is a ubiquitous hormone key to the global diabetes epidemic. Glucose is the body's fuel, and insulin the distributor of the fuel. Produced by cells in the pancreas, insulin encourages the conversion of glucose into its stored form, glycogen, or fat. When insulin levels are elevated, fat accumulates. When low, fat is broken down. This hormone lies at the heart of a worldwide health crisis.

For its part, insulin-like growth factor is the worm's simple version of the key hormone of human growth, which is secreted by the pituitary and ratcheted up during puberty. Both hormones made key signalers of the endocrine system, the body's intelligence network, controlling the release of many other hormones like adrenaline. Collating information from the immune and nervous systems, the endocrine system plays a vital role in growth, including the sexual changes that occur with puberty. It influences our metabolism and ensures that we have the proper levels of sodium and water in our bodies. The endocrine system, in short, is responsible for the stable internal environment that makes the body's crucial chemical reactions possible. The discovery that the Grim Reaper formed a part of a fundamental human pathway pushed the stakes for Sweet Sixteen, already high to begin, to spiral higher. "It was a race," Ruvkun recalled. Neither side was willing to lose.

The Race for Sweet Sixteen

The mechanism was nonlinear but logical. Downgrading the gene for insulin and insulin-like growth factor doubled the worm's life span. The gene specified a hormone receptor embedded in cell membranes. Because messenger hormones like insulin circulate in tiny amounts, they need boosters, or amplifiers. Receptors on the cell wall change shape when they come in contact with hormones. The outer part of the protein responds to the hormone; the inner part sends the signal to the cell. The mechanism resembles a relay race in which the messenger passes the baton to a recipient. The worms' cells become deaf to insulin signaling; they burn glucose more slowly. The fact that *daf-2* was a receptor and *age-1* a downstream runner meant they had found a molecular signaling pathway.

But why an insulin gene? First, Kenyon later wrote for the *Philosophical Transactions of the Royal Society*, the fact that system was triggered by hunger "was the first clear indication that genes encoding nutrient sensors regulate ageing. . . . Evolutionarily conserved hormones" controlled a "danger signal" that shifted the body between two states. If food is plentiful, the insulin receptors, which bind the insulin, promote energy storage, growth, and aging. If food is scarce, the lowered insulin signal shifts the cells to a state of maintenance, repair, and long life.

The question was how. The key to answering that was to clone the long-life gene. Ruvkun was using one cloning method and Kenyon another. Two years passed as the Kenyon lab's technician, Jenny Dorman, and then later, Kui Lin, a talented postdoc, worked on finding the gene. Dorman had already shown that Sweet Sixteen was necessary for the life-extending effect of *age-1*.

In June 1997, at the International Worm Meeting in Madison, Wisconsin, the Ruvkun lab made a mistake. "They spilled their guts," a lab member recalled, announcing that Sweet Sixteen gene was a "forkhead," so named because, when it was discovered in flies, the mutant showed an odd head structure that looked like a fork. It was a famous class of genes important during development because malfunctions in it caused cancer and diabetes. At the same meeting, the Kenyon team had a poster showing its progress.

The two labs ran neck and neck. By late August, Kenyon's team had identified the gene and cloned it. She took the unusual step of writing Ruvkun to say they were submitting their report for publication. Ruvkun's team pulled two all-nighters in a row, rushing to get the paper ready for submission. On August 19, 1997, Kenyon sent their report to the journal *Science*. On August 20, Ruvkun rushed his to *Nature*.

Ruvkun's report appeared first, in late October, with postdoc Scott

Ogg as the lead author. Two weeks later, Kenyon's paper appeared in *Science*. Whereas Ruvkun wrote on diabetes, she focused on life span. Sweet Sixteen revealed an entirely "new genetic regulatory program that extends youthfulness and postpones death." Both confirmed that a tiny snippet of DNA directed an integrated circuit that extended life in hard times. When Sweet Sixteen turns on, it produces a protein called a transcription factor that moves from the cytoplasm into the nucleus, where it encases DNA. The transcription factor made by Sweet Sixteen programmed cells for longevity.

It was a "turning point," said Heidi Tissenbaum, "the most amazing proof that tweaking a single snip of DNA could regulate so many life processes." The Ruvkun *daf-2* insulin paper had become "one of the most cited papers in the genetics of aging," wrote Tom Johnson. Kenyon, whose 1993 paper has been cited more than a thousand times, received the popular credit for being a pioneer in the field of longevity. For years afterward, Ruvkun resented Kenyon's aggressive push into his research area. She, in turn, felt the resentment was unfair. "We're a dysfunctional family," he said to me when I first visited his lab in the spring of 2002. At just the moment the world suddenly showed interest in their arcane research, the old idealism of the worm world disappeared.

Sex and Signals

Now that they knew the signal was hormonal, the insulin gene discovery set off "an explosion of analysis" wrote the NIA's Huber Warner. In France and the United States, researchers looked at mice with reduced insulin signaling. In England and the United States, they looked at flies. An Israeli doctor and two Hawaiian demographers began looking in centenarian populations for human longevity genes. The Ruvkun lab turned to the worm's nervous system.

By accident, Kenyon was soon to return to the question of sex and life span. Evolutionary biologists assumed there must be a trade-off between reproduction and longevity. Animals that die young generally reproduce fast and often. Longer-lived animals reproduce less and later. In humans, evidence based on English parish records suggested that women who gave birth later in life lived longer than those who did so earlier. It appears that eunuchs outlive normal men. (In the 1930s at a Kansas mental asylum, sterilized male patients outlived their guards and doctors.)

The molecular geneticists of aging disliked the determinism of the trade-off theory. "[Evolutionary biologists] are constantly telling you what you can't think," Ruvkun once said. In Kenyon's 1993 experiment, Jeanne Chang removed the reproductive systems of both normal worms and *daf-2* mutants, and found that it did not extend life span of the normal worms, showing that lowered fertility did not cause their longer life. Something odd happened, however, that they did not report publicly. The *daf-2* mutants' life spans, already double that of normal animals, extended still further, but they did not know how long, because her lab rotation ended and the worms still lived.

Cynthia Kenyon was anxious to explore the mystery of sex and longevity. The way to do that was to neuter the long-lived mutants to see if they lived still longer. Again, no one in her lab would do the experiment. For several months, however, a home-schooled twelve-year-old girl named Honor Hsin had been attending her developmental biology lectures. One day, Hsin asked Kenyon if she could work in her lab. Kenyon balked, but then she had an idea: let Honor Hsin try the reproduction experiment. The task was difficult, but Honor seemed very passionate about science.

The second puzzle was to understand the signal's scope. How was it sent, and from where? Did the insulin receptor mutation have to be in all tissues, or in a couple, or only one? To answer these questions, Kenyon wanted to apply a technique of classical genetics called mosaics. Mosaics, from the ancient Greek *muse*, are animals with mutations only in selected tissues, say the skin or nervous system, thus mixing traits in a single animal. Mosaics are best known in intersex individuals, like the fictional main character in Jeffrey Eugenides's novel *Middlesex*. In a lab, mosaics offered key tools to analyze a trait's origin.

Two years earlier, a tall, gangly, twenty-one-year-old Argentinean named Javier Apfeld had signed up for a rotation in Kenyon's lab. Apfeld had trained in chemistry in Buenos Aires and in microbiology at MIT. At a departmental party, talking to Kenyon, he got hooked on the enormity of the aging question. He read every study he could, including the accounts of Pope Innocent VII's 1492 attempt to revive himself on his death bed with blood transfusions from young men.

Apfeld set to work, creating worms that lacked the *daf*-2 receptor in specific tissues, such as the skin, intestine, muscle, or nerves. He sorted through tens of thousands of generations. He spent months watching as the worms were born, matured, mated (or, more often, fertilized themselves), grew old, and died. His three-month rotation in Kenyon's lab ended, but he stayed on. Two years passed. He kept going.

In January 1998, he noticed a plate with a worm that should have been dead. By early February, almost sixty days had passed. When he checked, he saw that only a few of its cells in the skin lacked the *daf-2* receptor. By March, he confirmed that worms with *daf-2* removed from much of the skin had double the life span of the normal worms. If it was taken from the nerves in one mosaic, their life span also doubled. If it was taken from the intestines, their life span increased by 30 percent. The experiments suggested the gut, nerves, and skin sent the longevity signal to the rest of the body.

At the same time, Honor Hsin worked with a laser microbeam on the experiment. She lasered the precursors of the seed cells, the sperm and the egg, of an infant worm less than 0.1 millimeters long without killing the animals. The four cells lined up in a row of four circles. The outer two circles formed the casing of the ovaries and uterus or sperm; the inner two formed the precursors of the actual egg and sperm. In a narrow microscope room lined with books, one had to fire the microlaser with a foot pedal, aiming at the two inner circles. You set the laser on its highest power and ratcheted the intensity down. In the microscope the shot made an incandescent blast. It was incredibly difficult work.

Removing the entire gonad did not further extend the life span of normal worms, so there was no trade-off between reproduction and longevity. But if she removed just the precursor seed cells from normal worms, leaving the casing, their life span increased by 60 percent. It seemed like a body with an empty gonad kept waiting for its seed cells to send a signal to mature. If she removed just the infant seed cells from long-lived mutants, life span increased an astonishing fourfold. The seed cells must send another signal, they decided, separate from the insulin signal. They thought it was sent by a steroid hormone, because one steroid defective mutant did not live longer than normal. When the two interventions combined, they extended healthful life by four times the normal span. It took their breath away. "It was like God was allowing us to see something we weren't meant to see," Kenyon said.

The "Book of Life"

Now dozens of researchers were undertaking longevity gene searches in flies, mice, rats, and rhesus monkeys. Heidi Tissenbaum joined Guarente's lab as a postdoctoral associate, and "Lenny really begged me to ask if sirtuins [the proteins produced by *SIR* could extend life in nematodes," she recalled. She devised an unbiased screen to test the idea. Tissenbaum found that adding extra copies of the worm *sir-2.1* gene, its version of *SIR2*, appeared to extend life span up to 50 percent. The extension required *daf-16*, seeming to unite the two longevity gene pathways.

In 1999, the biotechnology industry, fueled by a soaring stock market, ran hot. "Angel investors," people who in former years had staked Broadway shows, began to look at longevity gene research. It is hard to remember now that in the early years, many academics objected to the bipartisan effort linking government grants, university discoveries, and business profits (the bill sponsored by Birch Bayh, D-IN, and Bob Dole, R-KS). Kenyon had tried to form a longevity company but failed. Now Guarente contacted her. Arch Ventures, a Chicago investment firm, and its biology specialist, Cindy Bayley, had advised him to call Kenyon.

When Apfeld's paper on the location of the insulin signals appeared in *Cell*, and then Honor Hsin's study on the reproductive system and life span was published in *Nature* (Hsin was fifteen), Cynthia Kenyon's ideas took off. She lectured to packed auditoriums in Lisbon and Beijing and was invited to speak at Harvard and Yale. She won the King Faisal International Prize for Science, along with gene hunter Craig Venter. She was profiled in *Glamour* and the *New York Times*, interviewed on television by Alan Alda, and gave talks in the Netherlands and England and lectures at the University of Washington. Promoted to full professor, she was named to the prestigious Herbert Boyer Distinguished Chair. She went into her department chair's office and argued for a raise. Her Noe Valley clapboard house was a retreat from the growing buzz. On some nights, Cynthia Kenyon pulled out her six-foot-long Alpine horn and blasted a shot to the world.

On June 26, 2000, the sequence of the human genome hit the newsstands with great fanfare. At the White House, the rhetoric was as hot as the weather outside. President Bill Clinton and Prime Minister Tony Blair simultaneously announced the discovery of the genome "book of life." The discovery offered a "revolution in medical sciences whose impact can far surpass the discovery of antibiotics." Comparing the map of every human gene—twenty-one thousand or so at the time—to the triumph to the map of America produced by Lewis and Clark, Clinton added, "Without a doubt, this is the most important, most wondrous map ever produced by humankind . . . [an] instruction book previously known only to God."

Few in the White House rotunda noted that the Human Genome Project depended directly on the original work done on worms. A few magazine cover stories included sidebars on the precursor worm genome completed in 1998, when it became the first multicellular organism to have its DNA listed and read.

Fewer still noticed the worm researchers' efforts to understand longevity genes. But one audience was watching closely: venture investors with money to burn.

5 *Money to Burn: 2000–2003*

In 2000, a turning point was reached in genetics. It became cheaper to process genetic data than it was to understand it, and numerous biotechs started up to fill the gap, selling "hits" or promising molecules that could intervene in gene pathways. The year opened a "biotech century . . . remaking the world," wrote the Foundation on Economic Trends president Jeremy Rifkin. The information and genetics revolutions, fusing with a global economy, would apply gene discoveries to problems in human health. Pioneer companies like San Francisco's Genentech, the highly profitable manufacturer of synthetic insulin in the 1980s, pointed the way. Books like *Rapture: How Biotechnology Can Save the World* and *Bioevolution: From Alchemy to IPO* promoted the vision. A headline in the *New York Times* could spark a gold rush.

Shortly thereafter, I entered the story, traveling to Boston and San Francisco to Kenyon's lab. I became caught up in the romance of Kenyon and Guarente's company, Elixir. Their cofounder, forty-eight-year-old Cindy Bayley, loved whaling adventures and saw Elixir as a twenty-first-century version of the *Pequod* in *Moby Dick*. A wiry-haired, voluble risk-taker, Bayley raised the capital and outfitted the vessel. She hired the officers and negotiated their percentages based on their duties. The investors provided the tools. Like the mismatched officers of Melville's ship, the scientists came together in cramped quarters "to harness a natural resource," Bayley said to me, "with a huge payoff if they succeed."

They hired as chief science officer Pete Distefano, a Washington University biologist, editor of the *Journal of Biological Chemistry*, and Millennium Pharmaceuticals renegade. As they made the rounds of investors, they explained the significance of the gene discoveries. Kenyon's longevity pathway was triggered by a receptor resembling the human insulin and insulin-like growth factor receptor, causing life-extending changes in flies and other lab animals. Guarente's *SIR* gene was a triggered by a common coenzyme. Both genes seemed to operate in the same pathway in the worm. The company had expertise, a willingness to gamble, and an overheated economy flush with dot-com fortunes.

Other researchers followed suit, so many that *Science* magazine created an online "knowledge environment" to mimic the old *Worm Breeder's Gazette* and share longevity discoveries from many fields immediately. The Science of Aging Knowledge Environment (SAGEKE) sought to jump start interdisciplinary research. A new term, "gerotech," described the gold rush of science rivals suddenly competing for investors. The Boston company Centagenetix used data from its human centenarian project to seek life-extending gene mutations. A Montreal company called Chronogen had a different long-life gene. LifeGen, founded by Rick Weindruch and Tomas Prolla in Madison, Wisconsin, used information from the thirty-year-long rhesus monkey study on caloric restriction, to identify markers or signposts for aging—decreasing muscle mass, declining heart, and artery health—and running DNA through a new technology called GeneChip, a small plastic chip that could tell which genes were expressed at any given moment in life. The companies Eukarion and Rejuvenon followed more traditional drug-search protocols, while more adventurous, out-there start-ups Juvenon and Health Span Sciences sought to sell antiaging food supplements. Of these, the company with the most credibility remained Geron, working on a synthetic version of the enzyme protector of chromosome endings, and on the health preservation potential of stem cells.

Altogether, the newer longevity companies raised more than $100 million among them. The dream may have been social improvement, but the fuel was money and the need for profit. Gene patents protecting the study techniques of a gene for the length of the patent made the new industry possible. New tools of automated genetics sequencing and Internet communication made it viable. Previously, aging study had been "just grinding up rats and looking at their arteries," said the Lawrence Berkeley laboratory biologist Judith Campisi, who explored the relation of cancer and aging. Once a circle of friends, lab researchers raised in the '60s and '70s, now became a field of aggressive rivals. The qualities of their science matched the qualities of the era: to voyage, into the cell; to explore new regions, of DNA; to raise new money and to break limits, of previous human experience.

The companies offered similar business plans. Find a promising molecular longevity pathway. Seek compounds to tweak that pathway. Test the compounds in model organisms. Identify people who know how to make drugs. Finally, find a disease to treat, because the FDA was never going to approve trials for a drug to cure aging.

Elixir began in East Cambridge, Massachusetts, a neighborhood of brightly painted wooden row houses, student housing, old factories, Greek diners, and abandoned lots. It was snowing the first time I visited the tiny office sandwiched between an elderly lawyer who walked around in slippers and a dying dot-com called Dynamic Ideas. A whiteboard on the floor showed in colored marker the steps in Kenyon's and Guarente's gene pathways. The new company identified chemical compounds that might reduce *daf-2* activity. Two of these compounds had been tested already by Merck Pharmaceuticals for diabetes, so the compounds had been shown to be safe. The Elixir plan was to run high-speed tests to explore the biochemistry for drug targets.

Prospective investors filed in monthly. Some looked macho and tanned in tight golf shirts; others were rumpled academics well paid to evaluate their peers. The company would use patents to build "picket fences" around their technology and ideas. These fences only needed to hold long enough until they could patent a few compounds and sell them for millions. The company's PowerPoint diagrams glowed on the screen like modern versions of the maps of the first European explorers, their targets on twisting threads of brightly colored genes and proteins. Although it cost roughly a billion dollars to develop a drug, the market for a blockbuster could reach $40 billion. For their part, universities opened technology transfer departments complete with intellectual property lawyers, seeking to capitalize on their faculty's best ideas.

Not everyone liked the new cozy relationship between business and academic science. "What we're talking about is the influence of money on research that my journal and other journals publish. The distorting influence of it," said Drummond Rennie, deputy editor of the *Journal of the American Medical Association*, in 2005, to science investigative writer Daniel Greenberg, "it's like a kick in the testicles."

No matter how many PowerPoint slides they showed, in 2001 the longevity gene idea as a serious business investment did seem a bit preposterous. The key was to find the analogous antiaging interventions in mammals. Several European and American labs raced to do just that.

"It Worked in Mice"

Near a hidden courtyard just off the busy Parisian thoroughfare of the Rue de Tolbiac, all through 2001 and 2002, the French National Institute of Health and Medical Research assistant professor Martin Holzenberger watched his prized mice. They were prized, because they had only one copy of the insulin-like growth factor receptor instead of the normal two. By the end of the fall of 2001, he discovered that inhibiting the insulin gene pathway could extend the lives of female mice by some 33 percent, the males by 16 percent. The animals resisted oxidative stress, remained free from cancer, and responded to smaller doses of insulin than normal. They bore a number of similarities to mice on caloric restriction. Then Harvard's Ronald Kahn discovered that deleting insulin receptors from fat cells in fat insulin receptor knockout (FIRKO) mice extended the life spans of both males and females by an average of 25 percent. Research accelerated.

Mice posed several challenges to aging research, however. They possess separate genes for the insulin and insulin-like growth factor receptors, so the experiments became more complicated than those on flies and worms. Second, mice live on average two years, so experiments took a longer time. Third, the dustbin of medical breakthroughs bears the label "It Worked In Mice."

Still, for three years the Southern Illinois University biologist Andrzéj Bartke had run experiments comparing the life spans of normal and Ames dwarf mice, which lack a pituitary gene with the spiritual name prophet of pituitary factor 1 (*PROP1*). Bartke struggled for grant money, because antiaging hucksters and even some medical experts extolled human growth hormone as a key longevity treatment. In 1996, he published his initial results in *Nature*. The dwarf mice lived 50 percent longer than normal. "For the first time," wrote journalist Greg Critser, "a single gene mutation in a mouse could be shown to extend life span, thus placing the mouse and mammals in the same arena of life-extending genes as the fruit fly, the worm, and yeast." Moreover, the Ames mouse, along with Snell dwarfs and GHR knockouts (Laron dwarf mice) showed reduced incidence and delayed onset of cancer and were protected from age-related insulin resistance. Two of the mutants, Ames and Laron mice, maintained youthful levels of cognitive function into very advanced age, and the Snell dwarfs delayed the aging of their immune systems.

For these reasons, the University of Michigan's Richard Miller called the Ames mouse "the single biggest longevity breakthrough in mammals to date, and the death blow to those who say aging is too complicated to study." In January 2003, Bartke made a more astounding longevity breakthrough. When his lab moved from Carbondale to Springfield, Illinois, he forgot one dwarf mouse with a disrupted growth hormone receptor gene. They left the mouse in Carbondale, where it passed away and was charted. Later, when checked its chart, they realized it had lived nearly five years, double the normal span.

Miller had a dwarf mouse he named Yoda, a *Pit1* mutant that was otherwise virtually identical to Ames dwarfs, which lived almost five years as well; Miller wrote a poignant memorial when his longest-lived animal died.

The triggers were growth hormone receptors. Human growth hormones increase bone density and muscle mass. Athletes abuse it because it is virtually untraceable in urine. But growth hormone is known to have many negative effects, such as raising the probability of cancer and other late-life diseases. Between ten and thirty times a day, your hypothalamus, a tiny brain region that controls body temperature, hunger, thirst, and other functions, signals the pituitary gland in the brain. Each time the pituitary gland receives the signal, it sends out a small amount of growth hormone, which latches onto growth hormone receptors, like antennae, that trigger the secretion of insulin-like growth factor (IGF-1), promoting growth until a child matures. The mouse and worm studies suggested that for healthful longevity, you wanted to limit growth hormone and steroid signaling.

Abundant evidence linked smaller size with longer life in mammals and even, to a lesser extent, in humans. Every veterinarian's office sports a poster showing that the larger your dog is, the shorter its life. Small dogs are natural *IGF1* mutants, much like some of the long-lived lab mice and worms. When researchers, funded in part by Purina, later showed that this gene mutation caused smaller dogs' long life spans, the discovery made the cover of *Nature*.

The human growth hormone business made a multimillion-dollar industry, though, and doctors pointed out that some of the dwarf mice could not reproduce or regulate their temperature (Yoda required a large female, Princess Leia, to keep him warm.)

Bartke responded by studying a different mutant that generated growth hormone but lacked the growth hormone receptor. This lived longer than normal but was fertile and hardy. Labs working on both flies and mice showed that a mildly reduced insulin-like growth factor pathway extends life span with virtually no effect on body size. "I try not to take a dogmatic position on either side," Bartke told an interviewer, "but the growth hormone ads are blatantly commercial and incorrect."

Armed with these results, Miller became a proselytizer in Washington DC, where the NIA devoted too much of its money, he felt, to research into Alzheimer's and heart disease. A Yale MD–PhD, Miller complained that the sizeable National Institutes of Health budget set aside less than 0.1 percent for the study of aging in lab animals. Here they had found genes with human counterparts that could attack many of the aging diseases at once, yet the biology of aging received a fraction of the money. "If you spent an iota on the biology of aging that is spent on Alzheimer's," said Miller, "there might be a huge reduction in the number of sufferers, and the disease's duration would be shorter and less traumatic."

Similar results in flies followed. The University of London's Linda Partridge and Brown University's Marc Tatar both announced the flies with an inhibited insulin pathway lived longer and seemingly remained healthy. Insulin played a "central role" in extending longevity, said Tatar. When the chemical messages sent by an insulin-like hormone are reduced inside the fat cells of a fruit fly, life span increases by an average of 50 percent. "Insulin is a key player in the aging process," Tatar concluded. "It should change the way we think about aging."

In his small office, Bartke tried to make sense of the conflicting discoveries by creating a flowchart of the various theories of aging, centered on decreased growth hormone, with arrows pointing to various decreased incidences of disease. At the center of the overlapping Venn diagrams was one little circle, labeled "increased insulin sensitivity." Long life is the body's natural response to lower insulin levels and correlates with improved late-life health. "The authors of the French study very much emphasized that you can have extended life without these costs," Bartke told the SAGEKE website. The discoveries in such different organisms hinted that the insulin signaling system evolved very early in life's history and the results spurred the research into human versions of the insulin genes. "We have a glimpse of the control panel by which aging can be regulated," concluded Miller, noting there was now "incontrovertible evidence that aging can be slowed in mammals, fairly easily."

The discoveries "narrowed the field," wrote the NIA's Huber Warner, who heeded Miller's pleas to create a research grant series based on the insulin pathway. "If it's true in flies and mice," said Cynthia Kenyon, "the chance that it is not true in humans is small."

Twenty-Two Labs

The lab battles intensified. By the summer of 2002, Kenyon's dark, narrow, labyrinthine lab at the top of Mount Parnassus in the Health Sciences building was unbelievably busy. It was here that Reno native Andrew Dillin discovered that the insulin pathway increased life span *only* when tweaked in adulthood. The Kansas-raised Coleen Murphy undertook a huge screen of all the genes controlled by the master superintendent *daf-16*. Jim Thomas's University of Washington lab was doing the same experiment. The Kenyon lab published a paper in *Nature Genetics* exploring the different inputs on Sweet Sixteen, including insulin, growth, steroid, and reproductive signaling. "We counted twenty-two labs now working on aging genes in *C. elegans*," said Murphy.

The teams used a new scientific tool, RNA interference (RNAi), featured in a 2002 *New York Times* profile of Gary Ruvkun highlighting his pioneering use of it to understand timing genes in the worm. RNAi controlled gene activity by interfering with the messenger, ribonucleic acid (RNA). Ribonucleic acid takes the DNA template and transports it to the ribosome, the address in the cell where the proteins that do the work of genes are produced. If you disrupt the transporter, you stop the gene activity. RNA had an important natural role in defending cells against viruses and in directing growth. The Kenyon and Ruvkun labs used RNAi to screen for every gene that might extend life.

As the discoveries accumulated, researchers clashed with evolutionary theorists. "People who think they are going to find a fountain of youth, whether at the molecular or any other level, are not going to be successful," the well-respected evolutionary biologist George Williams told the *New York Times* in 2002. "You have to look at the trade-offs. The reason a gene may prolong life is because it does other things that are deleterious. Evolution has optimized the distribution of genes over a life span. If there is some rare gene that retards senescence, it will have other effects that are not worth the benefit."

The FDA would never approve of a large-scale trial of a treatment for a something no one considered a disease. As the scientists and officers of the new companies made the rounds of the same venture investors, they argued on behalf of "biomarkers," medical yardsticks, such as thinning bones or hardening arteries, as measures of aging. "Look around you," Guarente once exclaimed at the wrinkled faces and balding heads at one such meeting, "everywhere, signs of aging!"

Their initial business model was that of statin drugs. More than a century earlier, a German pathologist discovered cholesterol was responsible for the thickening of the arterial walls. In the 1950s, this discovery and the implications of it were later observed in a Framingham, Massachusetts heart study showing a high correlation between high blood cholesterol levels and coronary disease. In the 1970s, a Japanese microbiologist accidentally discovered a liver enzyme that inhibited cholesterol biosynthesis. Merck Pharmaceutical researchers uncovered a natural version of the enzyme called lovastatin, the first member of the statin class of drugs. Statins were safe, hugely profitable, and nearly unanimously acclaimed, dubbed by the *New York Times* as a "miracle drug."

Few in those first days talked much about the ethics of selling a drug to extend healthful life. Would it only be for those who could afford it? One idea was that it would be prescribed exactly as statins were, to patients at risk. But we were all at risk for aging. How would you conduct a clinical trial? How get the FDA to approve it?

To address such dilemmas, some pointed to the evolution of life-span differences that occurs naturally. Different mechanisms could extend life span in the wild. It could be size; large size generally means longer life, as with sequoia trees, elephants and whales. It could be wings, as with long-lived parrots. The wings meant bats lived much longer (thirty to sixty years) than their cousin mice (two years). It could be intelligence, loosely speaking, as with humans and chimpanzees, which averaged thirty-five years of life. If you master your environment by any means—intelligence, size, flight—eventually life span will extend naturally, and sometimes radically. Because it seemed the key genes were so few, organisms could develop different life spans in just a few generations, "an evolutionary blink," Miller wrote. "Nature discovered the backbone once. It discovers longevity over and over."

Such philosophical questions became more pressing as the first conference of the Molecular Genetics of Aging in four years gathered over a weekend in October 2002 at the Cold Spring Harbor Laboratory on Long Island Sound, where a new voice disrupted the show.

"Reverie"

The Molecular Genetics of Aging conference was the major conference of the new field, and participants included fly, mouse, yeast, and worm researchers; gerontologists; and representatives of some of the longevity companies. The field's elder, George Martin, tall and grandfatherly in a southwestern sweater vest, offered support. Even James Watson, then Cold Spring Harbor's director, made it down.

At breakfast on Saturday morning on the sun-lit terrace overlooking the bay, newcomers crowded around Guarente and Kenyon. Investors and new graduate students trolled the patio, seeking key researchers like Stephen Helfand, discoverer of a fly gene called INDY (after the Monty Python movie line, "I'm Not Dead Yet"). The Swedish postdoc Malene Hansen, who was working in Kenyon's lab, told me, "I'm new to the field, and you hear all these stories. I've read the papers and then to see these people, it's really exciting." At lunch, scientists compared their deals and discussed discoveries. In tight jeans, black T-shirt and an earring, a UCLA fly researcher told Elixir's Cindy Bayley, "I have two genes. They're certainly," he paused, "better than INDY."

"You need patent protection," she said, handing him a card. "You have to get the protection before you publish."

"I put it out in the announcement for this conference."

"That will kill the European rights. But American rights are where the money is. You still have a year to secure your patent."

Late Friday, Gary Ruvkun had shown up in the bar, just back from a Buck Institute gathering in Marin County. The Buck Institute began in 1975 with the passing of Beryl Buck, widow of Dr. Leonard Buck, a pathologist by training who had inherited a sizable fortune from his rancher and oil man father, Frank Buck. The couple had founded the first of several funded centers for the study of the biology of aging in a new partnership between government and business. At the time one of the only institute dedicated solely to aging research, the Buck Institute received some of its biggest infusions of cash from the Ellison Medical Foundation founded by Oracle's Larry Ellison. Ellison was an avid amateur microbiologist notorious for exploring longevity cures.

In an automotive machinist's blue T-shirt and baggy gray pants, wearing his name tag on his belt, Ruvkun told a group gathered around him about his first attempt to join NASA's search for extraterrestrial life by proposing to put a polymerase chain reaction machine on Mars, the kind of a machine used in restoring DNA at crime sites. "They turned me down," he said. "But I'm keeping after them."

Guarente, in sandals and Bermuda shorts, chaired the first session on model organisms. He worried about the dangers of founding a company in limiting his own freedom in the lab. He would have to sign away the rights to his own discoveries. He might have to slow publication of his lab's findings. He might not see his discoveries pursued. "I have no idea in the end," he later wrote, "whether the two ventures of business and science will be compatible."

On Saturday, a new voice took the stage. Guarente's former postdoc, David Sinclair, stood up to announce that he did not believe his mentor's discovery about the metabolic enzyme NAD being the link between his gene and nutrients. Rather, Sinclair had discovered a new longevity gene, *PNC1*, that made a "master controller of aging." The audience groaned loudly at the hubris. He would never forget it, nor they him.

On Saturday night, the arguments died down as everyone gathered for drinks to hear a poetry reading and jazz session on Long Island Sound. On picnic benches and in the back by the beer kegs, the new field's main figures gathered: Guarente, Shin Imai and his wife, Heidi Tissenbaum, Cynthia Kenyon's lab worker Andy Dillin, David Sinclair, and others. Townspeople came with picnic baskets. Families spread their blankets on the lawn to hear poet Donald Axinn read a poem called "Reverie." "Nothing is new, everything is new / You must know the players and translate their languages." A saxophone played. Wine glasses clinked. The first stars peeked out in the night sky. It was not hard to imagine Gatsby standing on the deck, arms raised, watching.

As I stood listening to the jazz on the Sound, I recalled something Kenyon said when I visited her lab that summer. "I have something in me that defies what I'm told," she said. "I'm not intimidated by the big names. I am willing to go after them."

In Washington

As the field gained momentum, the first discussion of the ethics of longevity research took place in Washington DC. In December 2002, the Bush Presidential Commission for the Study of Bioethical Issues turned from its tumultuous stem cell debate to entertain the moral and legal dimensions of longevity gene research. Chairing the Bush panel was the University of Chicago's Leon Kass, a Fellow in Social Thought at the American Enterprise Institute. In the journal *First Things*, he had written an essay titled "L'chaim and Its Limits: Why Not Immortality?" that brought him to the front of the conservative view of the longevity debate. "If it were up to us to set human life span," he asked, "Where should we set it, and why?" Citing sources from the Torah to Montaigne, Kass linked mortality with virtue and raised questions about the new sciences of stem cells, growth hormone replacement, and cloning. He praised Homer's Odysseus, who turned down the offer of immortality from the nymph Calypso. "The finitude of human life is a blessing for every individual, whether he knows it or not," Kass said.

Two research experts, Steven Austad and Jay Olshansky, presented the scientists' side. Austad was a respected University of Texas comparative zoologist and a contributor to *Natural History, Scientific American*, and *National Wildlife*. The University of Illinois demographer Olshansky was a well-regarded skeptic who coauthored a book criticizing scientific hype on longevity. The two had a famous bet based on Austad's claim that someone in their generation will live to be a hundred and fifty years old. Olshansky got so mad, he recalled, "We bet which of us would be alive when they were a hundred and fifty. We invested $150 each, and enlisted our children to administer it."

Austad described his work on the evolution of life span among wild opossums on Georgia's Sapelo Island, where the absence of predators led to longer-lived animals. Sapelo opossums lived on average 25 percent longer than their mainland cousins, with a 50 percent greater maximum life span. An English literature major before switching to science, Austad had loved islands ever since reading *Treasure Island*. Living in a trailer on Georgia's Sapelo Island, once a hideout for runaway slaves, he discovered that healthy life span changed naturally in the wild all the time. It could happen incredibly quickly, in a matter of a couple generations. There was nothing unnatural about exploring how it happened, he explained.

Then Austad turned to the looming aging crisis. "If current trends continue, the number of those over 65 will double by 2030," he said. "The number of centenarians will increase one hundred-fold in the United States. The Social Security system will be bankrupt by 2041. We face a looming societal train wreck if we extend life without improving the quality of health as we age."

Both scientists argued for more funding for longevity gene research in lab animals. "We have been wildly successful over the last fifteen years in extending the life spans of lab animals," Austad said. "The disparate processes of aging now appear to be connected. We can alter a single gene in a complex organism and by doing so, increase longevity and preserve functionality to an amazing extent." The goal of such research, he emphasized, "is not the prevention of death but the preservation of health."

To the scientists, however, the overall experience was disheartening. "It seemed like Kass did not understand," said Olshansky. "We're already doing this, improving health. We're trying to show how to do it better, less expensively, more efficiently, through the basic biology of Kenyon, Guarente, Ruvkun, and others."

As the year turned, Cynthia Kenyon took her discovery on the road, traveling first class with her new fame after she won the $200,000 King Faisal International Prize along with Craig Venter of the Human Genome Project. At the University of California, San Francisco, some $8 million from the Hilblom Foundation was donated to found a new Hilblom Center for the Biology of Aging, with Kenyon as director. It was to be part of a $1.5 billion Mission Bay campus that would double the research space of UCSF and mark the city's largest urban development since the creation of Golden Gate Park. The new lab was designed with meeting places for coffee to mimic the old Laboratory of Molecular Biology in Cambridge, England.

By the end of 2002, the mouse genome appeared in the journal *Nature*. Mice and humans, it was shown, each had some twenty-one to twenty-two thousand genes, and all but three hundred were the same. More studies were using mice to screen for drugs, because it was easy to "knock out," or remove, genes from mice. A Methuselah Mouse Prize, in homage to the famous eighteenth century Longitude Prize, spurred a research race for a longest-lived mouse.

Of course, mice were not perfect models. "What would really make sense," said Jay Olshansky, "would be a Methuselah Human Prize, not a Methuselah Mouse Prize."

As several labs committed to finding counterparts of the worm genes in humans, the rest of the world was not too concerned. At the end of his Washington presentation, Steven Austad voiced his skepticism: "Some of these long-lived mutants don't reproduce well," he said. They don't respond to cold. Slowing aging in fruit flies is very easy. If you decrease the temperature from 30 degrees Celsius to 18 degrees, you will increase their longevity by more than six-fold," Austad added. "Some of the really spectacular modern advances in increasing longevity . . . [are] only the refrigerator effect."

In 2003, if you had asked most biologists, few would have imagined finding a compound to influence the rate of human aging. Despite that reassurance, Olshansky concluded by warning the panel:

> Regardless of whether you and I want it to happen, the scientific research devoted to the issue of modifying the biological rate of aging will happen. . . . A number of scientists are creating companies where they intend to sell products to the public to influence the aging process, well-known scientists are doing this—I mean, people are going to battle against death. They are going to fight against aging, and developing these technologies are a fundamental part of the biological sciences. And it's going to happen whether we want it to or not.

6 *Longevity Noir: 2003–2004*

The water dripped down from the wall air conditioner units onto the lab benches. It could be perfectly sunny outside, and still it would drip. In the late spring of 2003 in the century-old stately Harvard Medical School Department of Pathology building, David Sinclair, age thirty-four, was a new assistant professor competing with some of the best minds in his field. His mother had been stricken by lung cancer some fifteen years earlier, and he felt an urgent need to do something with his post at a leading medical school. Sinclair was trying to trigger an enzyme with a plant compound. Not a mouse or worm enzyme, but one expressed by a human gene.

To do so, his collaborator had discovered two plant extracts that activated the human sirtuin enzyme. Now, Sinclair was testing the most promising of the two to see if it extended the life of a whole organism. On his desk, he had set up four dishes of yeast cells to treat with different concentrations of the compound. It was a difficult technique, and he did not allow anyone else in his small lab to work on the experiment.

One of the four trays had no compound. The other three had various concentrations of the same compound. He could see, day after day, that one dish was living longer and reproducing mightily. He was dying to find out which concentration was causing the effect. He even brought the trays home and worked at the dining room table. But it was a controlled, blind experiment. He could not look.

Sinclair had several other experiments going in his small lab. He studied another longevity gene called PNC1. At the Cold Spring Harbor Molecular Genetics of Aging conference the previous October, David Sinclair's claim that PNC1 was a "master regulator" of longevity elicited a huge groan from the audience. Such a claim was everything gerontology experts criticized about the new molecular genetics of aging. "We would tell him, 'David, don't say things like that,'" recalled the University of Wisconsin researcher Rozalyn Anderson, who was then a postdoc in his Harvard lab.

"Well," he would say a few months later, "people are taking me much more seriously now."

Few thought it was a viable strategy to attempt to stimulate an antiaging enzyme. Most pharmaceutical compounds did the opposite, interfering with a signaling pathway. True, nature could do it. The life-extending effects of caloric restriction needed to have some mechanism, and Sinclair's mentor Lenny Guarente had argued for years that the *SIR2* gene played a key part in the mechanism.

Sinclair had teamed up with a German-born plant expert named Konrad Howitz at the pharmaceutical manufacturer Biomol outside Philadelphia. Biomol made high-speed search tools for sirtuin inhibitors. People were interested in sirtuin inhibitors because Guarente, with other experts, had linked them to the p53 gene that suppresses the growth of cancerous tumors. Of course, since sirtuins turned p53 down, the idea was to deactivate them. Using his company's lab kit, Howitz had found a pair of chemicals that stimulated the human enzyme SIRT1, produced by the human version of the yeast gene. That was highly unusual, because the wisdom was that enzymes are difficult to stimulate, and so drug companies rarely looked for them. Howitz, an affable army enlisted man's son, was having trouble cloning the SIRT1 gene. He agreed to send Sinclair the Biomol testing kits in exchange for Sinclair's genetic material. "It was not a drug discovery motivation for us," Howitz recalled. "We were looking for tools we could sell."

The two compounds Howitz had found were polyphenols. Polyphenols were known antioxidants with many health benefits. They possessed multiple phenols, or carbolic acids—toxic, white crystalline solids with a sickly sweet, tarry odor. Many of the polyphenols had been found to have anticancer activity in the lab, and since they were present in some red wines and many olive oils it was thought that they might be responsible for the longevity benefits of the Mediterranean diet. Howitz's company had put to work its secret reagent called Develop and found more than a dozen phenols triggered SIRT1 in the lab. By the late spring they were studying all the plant polyphenols they "could get their hands on," said Howitz. Sinclair focused on one, called resveratrol, because it was a known compound with health benefits. Could it turn on an enzyme that extended life span?

The way to find out was to give resveratrol to yeast cells. He was dying to see which concentration was in the petri dish with the thriving yeast. Doing so might taint his experiment, but he could not stand the mystery for much longer.

"It's Hard to Keep Your Mouth Shut"

Born June 26, 1969, David Sinclair witnessed nature's power growing up in an exurb of Sydney, Australia. The loud hoots of laughing kookaburras, white and yellow sighing cockatoos, and rainbow lorikeets woke him every morning at five. "I spent almost every day in the bush. It was a new suburb, the animals hadn't quite figured out we were bad yet," he recalled. At the end of his block was a vacant lot, behind which the outback rose, a dusty, wind-blown menace.

Very early on he experienced a revelation. "At the age of three I thought of dying, of having my parents die or my pet cat, or myself, and I thought, aging is abhorrent. Why do we accept it? Death is horrible." David Sinclair loved art and animals and hanging around in his parents' laboratories, playing with paraffin and test tubes. They worked as industrial chemists in the medical diagnostics industry, and arguments over reagents ran around the dinner table. Their son put on their lab coats and goggles and played at being a scientist. The grandchild of a Holocaust survivor, in school he was a smallish boy with an unruly flop of brown hair and a smirk who once set off an explosion to see the reaction. Only when he went into his parents' labs did he calm down.

Sinclair was studying biology at Australia's equivalent of MIT, the University of New South Wales, when Kenyon's 1993 *Nature* paper captured his imagination. He felt a longevity gene search would not be as complicated as experts claimed. Yes, aging might be subject to thousands of individual variables, but "you did not have to worry about that," he reasoned. You did not have to understand everything about aging. All you needed was one intervention. "If we could just find one molecule to modify," he recalled, "we could have a major impact."

When he heard Guarente speak in Sydney, Sinclair felt even more galvanized as a doctoral candidate. "To postpone aging"—how could a scientist even say those words? Yet all around us, even in the life spans of our pets, rang the testimony that something controls the rate or quality of aging. Gerontologists did not believe it, but he loved challenging authority. He dogged Guarente to admit him to the MIT lab. Eventually Guarente agreed as a postdoctoral fellow.

He arrived in Massachusetts in early 1996. Sinclair and his friend Kevin Mills loved *Beavis and Butthead*, and the two young researchers taunted each other into working faster and better than either would have alone. The longevity story grabbed Sinclair's imagination but it made his backup plan. He did not attend science conferences as a rule, where he feared he would be clouded by other's ideas and might give away his secrets. "It's very hard to keep your mouth shut," he said.

He could trace the exact moment when his life changed—when he was nineteen and his mother was diagnosed with lung cancer. She was given a 5 percent chance to live. Though he disliked medicine's repetitive routine, he resolved to pursue research into the biology of cancer or health, hoping to transform his models of balsa wood and posterboard into actual genes and save the life of a woman he adored.

Then came his *Cell* paper on a mechanism of yeast aging and there was no turning back. Guarente recommended Sinclair for a Harvard assistant professorship. As a new faculty member, Sinclair visited Gary Ruvkun at Massachusetts General Hospital. Ruvkun told him that he "really wasn't into aging." Sinclair admired Ruvkun's coolness.

"I was lying, David," Ruvkun admitted.

Across the Charles River in 2003, Elixir researchers investigated several different gene pathways, but no single intervention showed a clear profit potential. They ran a massive, expensive scan of almost four hundred thousand artificial compounds, burning through hundreds of thousands of dollars, seeking a trigger of either the insulin or silencer genes or a blocker or "antagonist." The company had yet to decide between activating Kenyon and Guarente's genes as a way of lengthening life, or inhibiting them, as a way of fighting diseases like cancer. Pushed by its funders, Elixir held a critical meeting in June with pharmaceutical consultants including Merck's former director of research and development, Ben Shapiro, and its former chief scientific officer, Ed Scolnick. Scolnick and Shapiro repeated the mantra that the FDA would never allow clinical trials of a possibly dangerous compound to make people live longer. Researchers must forget about life span and go after moneymakers like diabetes and cancer. MPM Capital took control of the board, and Elixir merged with another Boston company searching for human longevity genes in centenarians.

In the Sinclair lab, some of the closest members of the team were kept in the dark, recalled Rozalyn Anderson. "The guy from Biomol kept calling, and a couple of others David let in on the secret." Sinclair stayed up late, taking the plates home and removing the daughter cells at the dining room table, monitoring the assays on the weekends. Finally, he could not stand it any longer. He had to know what concentration was in that one petri dish. He told his wife, "I've got to find out what A, B, and C [are] because C is living longer."

Late one night, alone, he snuck into his laboratory and opened the worn desk drawer, digging for the paper buried under plastic bags and tubes. It was a breach of protocol to see what was causing the yeast to live longer in the midst of an experiment. Cautiously, he read the page and saw that the longest-lived yeast buds lay on a plate with the smallest concentration. Resveratrol, from red wine, a health food supplement, stimulated a life-extending signaling pathway. "Imagine the Sir2 enzyme as a Pac-Man. They [resveratrol molecules] bind to and make it work faster," he told the PBS *Newshour*. Most drugs worked by mucking up a biochemical pathway. A tiny bit of resveratrol extended yeast life span apparently by fixing one. "Something very big," he told his wife, "has just happened."

"Throw Some of This on Your Flies"

Resveratrol is an organic substance derived from knotweed, peanuts, and the skin of red grapes distilled in red wines, particularly those from harsher climates. It can also be found in green grapes but is leached out in the processing of white wine. Under environmental stress, grapes produced the protein in their skin to protect the sugary, pulpy core. The compound tends to be strongest in pinot noir and loses its efficacy after the bottle is opened for more than a day.

Resveratrol had been the subject of several treatises claiming it had anticancer properties. Because the molecule was common and highly reactive, Elixir had not even included it in its massive scan of compounds. Few other companies would have even screened resveratrol. You could not patent it; it was already available over the counter at health food stores. It was a natural substance, whereas the huge, high-speed pharmaceutical assays tended to favor man-made, patentable, exotic compounds, preferably something already tested for another use. It seemed open-minded and smart science to look at something commonly overlooked because it is a botanical compound. Scientists have found cures in natural compounds many times, as with aspirin, known to Hippocrates as willow bark, or penicillin, discovered by Alexander Fleming in bread mold. The trouble was, resveratrol was considered a dirty compound, meaning it had so many different effects it was hard to control.

That summer, Brown fly biologist Marc Tatar visited Sinclair's lab by chance. When Tatar's scheduled Harvard appointment was canceled, he told his student, "Let's go visit David." Tatar was a lanky, kind, former high-school biology teacher famous for eating termites while on an early research trip to Kenya ("they tasted like unsalted butter," he said to students). He was editor of a new journal called *Aging Cell* and had created a reading club called the New England Biology of Aging Research Group to discuss new findings in an infant field. Few group members had the time to fight Boston traffic and meet up. David Sinclair was one of the few who attended.

Tatar caught his friend as he was on his way out the door. Sinclair called them back into the office, excited and restless. He had been working all summer on a compound, he said, that extended yeast life span. "It's amazing."

"What is it?"

"I can't tell you. But you can throw some of this stuff on some of your flies."

"Is it safe?"

"Oh, yeah, it's safe."

"Why can't you tell me what it is?"

"I've signed a confidentiality agreement."

After Sinclair and Howitz submitted the resveratrol paper to *Nature*, Howitz took off on a long-planned trip with Biomol brass to a sales meeting in Brussels, and then on a biking and barging vacation on the canals of Holland. The journal accepted and decided to fast-track their report. *Nature* still had the ethos of its longtime editor, Sir John Maddox, who pioneered the practice of putting in news leads in the magazine to hype reports. Sinclair could make a public announcement at the Growth Control in Development and Disease meeting in Arolla, one of the best science conferences in Europe.

Sinclair wondered why Konrad Howitz had taken off on vacation at this critical moment. All through the summer, the two had been planning how to word the announcement. What was he doing on a houseboat?

At Brown University, Tatar did not want his lab team working with a compound he knew nothing about, so he conducted the fly experiment himself. The first night he returned to Providence, Tatar carried the mystery powder in a Styrofoam box packed with ice. He had to mix it in with the fly agar, a gelatin-like media. To do that, he had to melt the agar. He put it in the microwave. It exploded. Finally, he melted the agar in hot water.

"What are you doing?" a lab member asked the following morning.

"An experiment."

"Uh-oh. Watch out!"

About a month and a half passed. Right before Sinclair left for Switzerland, Tatar called him.

"David! What the heck did you give me?" he asked.

"I can't tell you."

"Listen, my flies are living longer than normal. They're healthy. You have to tell me."

After much pressing Sinclair, vowing his friend to secrecy, finally confessed. "It's resveratrol from red wine."

"Red wine," Marc Tatar said. "Holy crap!"

Harvard prepared a press release. *Nature* placed an embargo on the announcement so that reporters were required to keep the news secret until the announcement was made publicly. This was a common technique to prevent leaks and spike a bit of sensationalism. Many of the science announcements in major newspapers are carefully written and edited weeks before scientists present their results. Professional journals compete ardently for the best biology papers and being a good, dramatic communicator was a significant advantage in getting published. *Nature* had a reputation for being keen on embargoes if it had a splashy paper. Resveratrol, reverse-it-all, looked to be as splashy as it gets.

In the Swiss village of Arolla, after Sinclair explained the discovery in an aging forum that Gary Ruvkun and Cynthia Kenyon attended, he was deluged with phone calls. The *Washington Post* put the news on page two. The *New York Times* put it on page seven. The Internet provider Comcast reported the discovery on its website, along with a graphic of a tipped wine glass. The fact that it came from red wine made a perfect touch. Not only did it appear to be one of the first scientific proofs of a natural compound to boost longevity, the holy grail since the beginning, but also an activator of a known life-extending enzyme. "I've been waiting for this all my life," Sinclair told the reporter Nicholas Wade of the *New York Times*.

Others in the field carefully studied and restudied the early online publication. It seemed to defy logic. Seasoned pharmacological researchers considered polyphenols to be problematic, because they had too many interactions and so gave off false-positive results in experiments. A basic tenet of any biochemical screen was always to validate a finding with a second, different type of test.

Investors, well-wishers, and a Hollywood actor interested in maintaining his youth began phoning Sinclair. Biomol put the news on its website, making claims about its role in the discovery. Sinclair and Howitz developed a theory to go with the discovery, which they called the xenohormesis hypothesis. "Hormesis" is the aging theory stating that a little bit of stress is good for you. Xenohormesis suggests that a fungus like yeast, growing on a plant, could signal environmental stress to the plant, inducing the synthesis of compounds that send the host plant into a state of repair, maintenance and longevity. "To a yeast cell growing on a grape," Sinclair wrote, "an increase in the resveratrol induced in the plant after injury or infection, might be a useful indicator that the food supply is about to become limiting." Sinclair sought to show that resveratrol, triggering the "activation of sirtuin enzymes," set off cell responses similar to those of caloric restriction.

Already, however, warning signs appeared. At Brown, Helfand found that overexpressing the gene increased fly life span, but he was accidentally overexpressing another gene in the experiment, so that fly conclusion was not tight. If he overexpressed the gene in the fly's eye, the cells died. Sinclair and Howitz had used a particular Biomol lab assay or test called Fluor de Lys that contained SIRT1 and a fluorescent marker. When you added resveratrol, the fluorescent marker was activated, but they had not done a control to test if resveratrol was merely affecting the marker. If they had, they would have found three hundred and three hundred molecules that were activators. Fluorometric assays were known to be prone to multiple kinds of false positives.

The public, however, loved the story. "The media drive the hype in our field. It's a big problem," Judith Campisi later told the online Science Network, itself one of the many new science media outlets serving a huge public hunger. Some researchers shun the attention. Sinclair returned reporters' phone calls immediately and gave good quotes. "It's as close to a miraculous molecule as you can find," he told *Science*. "One hundred years from now, people will be taking these molecules on a daily basis to prevent heart disease, stroke, and cancer."

Through the fall of 2003, Howitz and Sinclair sought financial backing. As Harvard and Biomol tried to patent the usage of resveratrol, Sinclair flew to Philadelphia to propose doing the patent together. It was an uncomfortable moment. They tossed around company names. Biomol cofounder Ira Taffer mentioned the word Sirtris. "That's a palindrome," said Howitz, latching onto it. Key DNA regulatory regions, turning genes on and off, often made palindromes of the letters A, T, C, and G.

Back in Boston, in September, Harvard dedicated a $260 million New Research Building, where David Sinclair would occupy a state-of-the-art corner lab. He talked with a longtime Harvard philanthropist, Paul Glenn, about a center for aging study. Sinclair was thinking about making his own business contacts to develop the discovery. He took a Harvard MBA course. The person he was most interested in was Christoph Westphal at Polaris Venture Partners, one of the largest Boston-area venture capital funds. Westphal, a self-described "serial entrepreneur," combined restlessness, skill, and showmanship. If he was interested, he would offer access to a level of funding unimagined by anyone in Philadelphia.

Tell Me Everything

In the fall of 2003, Westphal first visited Sinclair's new office. In the preceding four years, Westphal had cofounded five companies, two of which went public—Alnylam and Momenta. With German-born parents, Westphal spoke four languages and had an aggressive style, taking his companies public well before experts thought they were ready. He had been attracted to economics as an undergraduate major at Columbia University, which he completed in three years.

Westphal's parents began as doctors with little interest in business. His father, Heiner, went on to become a National Institute of Child Health geneticist. Christoph wanted to get the structural economic underpinnings of science the universities did not teach. "I was trying to figure out what to do in my life and my Mom said fine, first you go to medical school, and then you can figure out what you want to do with your life," he told several interviewers. He earned his MD from Harvard Medical School and PhD in genetics from Harvard University before going on to work in surgery in Cameroon. He came back from that experience determined to combine his business, medical, and basic science talents to form new companies to make drugs to help others. Charming, driven, and an accomplished cellist who loved chamber music, he made an adept orchestrator of the clashing egos in the newly emerging world of big finance and cutting-edge science.

When Sinclair and Westphal met, however, it did not go well, Sinclair recalled. "Tell me about these molecules you've found," Westphal said.

"I want you to sign a nondisclosure agreement first."

"I never sign those."

Sinclair balked. "David," he recalled Westphal saying, "If I walk out of this door, I'm not coming back. So I suggest you tell me as much as you can."

Sinclair began talking. Genes could control life span, many discoveries had proved that. He thought there was a special docking site on the sirtuin enzyme to which resveratrol attached.

Westphal had read the papers. Where other people might pursue a number of projects slowly, he preferred picking one and banking everything on it. If it failed, it was okay—most science failed—but he wanted it to fail fast. It did not matter if it was insulin or sirtuins, this biology could be a really important venture, more important than any other company he had created. It was a promising medical target. Sirtuins were more malleable than the insulin pathway. Resveratrol was a natural compound. God created it, if you wanted to put it that way. Most important, he noticed that the puckish scientist with an Australian accent made a great communicator of complicated ideas.

The initial meeting ended without a resolution. Sinclair said more than he wanted to reveal. Westphal was going to be busy the next several months with the start-up of another company called Acceleron. He promised to call when he had time. Sinclair, however, kept talking to investors.

The Shape Changer

Insulin remained a main focus of longevity gene research. In July, the Kenyon lab researcher Coleen Murphy published in *Nature* the discovery, also confirmed by Jim Thomas, that the protein made by *daf-16* choreographed a wide variety of life-lengthening genes, including antioxidants, stress responders, antimicrobial agents, and metabolic enhancers, while also turning off other life-shortening pathways. It did so by moving into the nucleus, where it bound to numerous receptors, as several labs had shown. A Kenyon lab paper in *Cell* by Natasha Libina was more specific about the precise tissues where the protein worked to extend life. A few other researchers were beginning to catch glimpses of the gene's human versions, called FOXO, in Ashkenazi Jewish and Hawaiian and Japanese centenarians. "It's like you're an archaeologist and you've found an ancient city, and it's so full of potential every time you turn around you make a new discovery," Kenyon later said. Speaking of Kenyon's insulin gene discoveries, UCLA prostate cancer researcher Charles Sawyers told the *Wall Street Journal* in October 2003, "Big pharma, small pharma, biotech and the cancer world are all over this [gene] pathway."

Now international students were tracking Kenyon down to work on aging in her lab. The first was Nuno Arantes-Oliveira, a middle-class Lisbon researcher and music buff who memorized every world capital and currency and collected every Beatles bootleg recording that ever existed. Arantes-Oliveira arrived at midnight at the airport. When Kenyon picked him up she talked so excitedly that she got lost three times on the way to her lab, where he was surprised to see people still working, though it was nearly 1 a.m. The next morning he had to give a talk on his research. Kenyon invited the lab group to her house on Diamond Street and offered him the position.

Arantes-Oliveira entered a world of intense creativity. "People were working fourteen hours a day. A fifteen-year-old girl was publishing in *Nature*," he recalled, "I thought, this is amazing." When he ran out of money during his first week, a postdoc wrote a personal check for the balance of his account. Arantes-Oliveira set himself a goal of combining the reproductive and insulin pathways. He removed the seed cells from infant worms and then decreased *daf-2* signaling when they became adults. These worms lived six times their normal life span. Two of his favorite mutants, which he nicknamed Paul and Ringo for the two surviving members of his favorite band, lived an astonishing ten times their normal life span. Marc Tatar compared the result to "pulling a rabbit out of a hat." Linda Partridge termed it "magic."

Later in the spring of 2004, Sinclair and Westphal met again and decided to team up. They began making the rounds of venture capital investors. They need not make any claims about aging, Westphal felt. They could attack the diseases of aging like cancer, ALS, heart disease, and Huntington's. It made an enormous amount of sense to make a huge bet on one intervention. They could have a new way of treating the major killers of Western society. In a few years, they should have multiple drugs. How quickly, how many, and which made the best indications were open to question. They were not interested in extending life span. They wanted to attack disease. They "really didn't understand just how powerful that idea would be," Westphal later told an investment newsletter.

Westphal and Sinclair incorporated their company and called it Sirtris. They raised money from seed funders and angel investors, including Joseph Kennedy III, with the bulk coming from Westphal's Polaris Ventures. Their company signed up award-winning researchers to its scientific board, like MIT Nobel Prize winner Phillip Sharp; Bob Langer, another MIT professor and a prolific inventor; and Thomas Salzmann, a former executive vice president of research for Merck. But they did not include Howitz.

"I asked Christoph many times if Konrad could be involved," Sinclair recalled. "I thought he'd be a huge asset. But I was no longer running the show."

They focused on one intervention, sirtuins. Could they lower insulin signaling by targeting sirtuins? They thought they could. Could they get a drug approved? There was no answer for that yet, but they would know in the next couple years. Many antiaging interventions like stem cells and tissue regeneration sounded incredibly sexy, but such innovations seemed to be half a century away. A sirtuin activator offered the closest thing to a legitimate blockbuster. Westphal took the unprecedented step of leaving his highly profitable venture capital position to join the company as its full-time CEO.

The debate about a looming aging crisis was also intensifying. A 2004 federal study showed the American health-care system was unprepared for millions of aging baby boomers. The Social Security system would be bankrupt by 2041. In Japan and Europe, the median age was projected to rise from thirty-six today to fifty-two by the middle of the century. By 2050, fifteen million Americans were projected to suffer from Alzheimer's, according to one study, at a cost of some $1 trillion. Aging in Europe exacerbated the economic crisis in countries like Greece, Portugal, Spain, and Ireland. The situation was worse in Japan, China, and Korea. In China, an unintended effect of the one-child policy was an aging population that lacked children to care for elderly parents. The world faced a potential "clash of generations," and policymakers argued about the best way to address it, in much the same way as they had a generation earlier when the National Institute on Aging was founded. Several countries now created their own such research institutes.

Scholars in a new field called age studies argued for an opposing scenario, in which older people provide support to younger people, work longer, and volunteer more than they already do. Studies at Yale, the University of Chicago, and the Rand Corporation show that a long-lived healthy society generated more wealth than a short-lived society. Healthful longevity meant better intergenerational contact and transfer of knowledge: children get to know their grandparents or great grandparents, volunteerism would rise, and new laws protecting the rights of seniors allowed people to work later. For these reasons, a longer-lived society experiences a series of self-reinforcing benefits, according to the University of California researcher James Carey. Longevity reduced the size of families and increased the contribution of grandparents, increasing wealth.

At Elixir, the company's efforts were scattered. In part, it sought to deactivate the appetite trigger called ghrelin. Toward that end, a commonly prescribed diabetes drug, metformin, looked enticing. It increased insulin sensitivity and improved mouse longevity by 38 percent by deactivating the receptor of the ghrelin hormone that monitored and insulin levels and triggered appetite. The problem was that metformin was generic. The company also tested a sirtuin inhibitor to treat Huntington's disease.

At the same time, Kenyon's former worker Javier Apfeld, then an Elixir scientist, discovered a "controller of a controller," the enzyme called AMP-kinase. AMP-activated protein kinase (AMPK), functions as a "fuel gauge" to monitor cellular energy status. It was a master metabolic regulator with great potential, if it could be druggable. On the other side of the insulin pathway the company began negotiations to in-license, that is, to get the American rights to a Japanese drug called mitiglinide that increased insulin production. The question was whether Japanese trials would suffice for the FDA, and whether this opposing approach would make a profit for a company torn in different directions.

Sirtris faced huge challenges as well. It kept a low profile at first. Even though they had raised millions of dollars, they had little in the way of solid science data to show venture capital investors. "When we first went to investors, we had no compounds," recalled Michelle Dipp, a talented Wellcome Trust former surgeon with a PhD in physiology on loan from the company in England. "All we had were unvalidated targets and David's yeast data." She was not sure how long she would continue with the assignment.

There were bigger problems looming. The former Guarente lab members, Matt Kaeberlein and Brian Kennedy, now University of Washington assistant professors, were working on two different claims made by Sinclair and Guarente and could not confirm either. The first was the claim that the *SIR* gene worked through the pathway of caloric restriction. Their experiment in yeast showed just the opposite—if you removed the gene and restricted the food, the yeast still lived longer.

The second problem was more serious. Kaeberlein and Kennedy, and two other labs, including the researchers at Elixir, were trying to show that resveratrol triggered the *SIR* gene in yeast, and it was not working. Kaeberlein recalled their attempts to contact their old lab associate about the discrepancy:

> We tried everything, different temperatures, different conditions, nothing. We e-mailed David to ask what was wrong. We invited him to send someone to the lab to show us how to do it. We offered to come to his lab. We got the distinct impression he was putting us off, saying he was too busy teaching so we decided to go ahead on our own.

Kaeberlein submitted the report questioning *SIR*'s effect on the caloric restriction pathway to *Nature*.

Earlier that year, *Science* had published a story about the rift between Sinclair and Guarente called "Aging Research's Family Feud." It appeared a little like David Sinclair was now a Harvard professor going after his former prof. "They're doing exactly what we're doing," Sinclair was quoted as saying of Guarente's company Elixir, "and it's a race." The story expanded on Sinclair's Cold Spring Harbor presentation that depicted PNC1 as the master controller, refuting Guarente's model citing NAD as the Sir-sensing mechanism. It featured Guarente shoveling his walk, pausing to say, "This thing has put me [through] so many emotions, some of which I didn't know I had."

At the sun-drenched shores of Hersonnissos, Crete, in late April, 2004, Brian Kennedy gave a further broadside, explaining why he and Kaeberlein thought the *SIR* gene did not work through the caloric restriction pathway. After speaking in the gleaming auditorium where researchers from around the world the latest findings against the backdrop of Minoan civilization, Kennedy collared his former professor about the *Science* story. "I never said that!" Guarente exclaimed.

Changes in science communication impacted the idea's reception and the backlash. When *Nature* rejected Kaeberlein and Kennedy's report on the Sir2 independence of caloric restriction, they published it in a new web journal founded by Nobel winner Harold Varmus called the *Public Library of Science*. "It was good *Nature* rejected it," recalled Kaeberlein. "The result was everyone in the world could read it for free." The Kennedy and Kaeberlein experiment showing resveratrol did not affect yeast Sir2 appeared in the well-respected peer-reviewed *Journal of Biological Chemistry*. Of course, one of the labs with conflicting results belonged to rival company Elixir, which did the chemistry. Still, the failure to replicate Sinclair's results made a problem that festered in e-mails, journal letters-to-the-editors, and conference gossip. "It's a big thing. There has to be a commitment to truth," observed Kenyon of the resveratrol claims. "Scientists are trained to be very objective. You can't allow a bias, no matter how much money you've raised."

In the autumn of 2004, at the next idyllic Cold Spring Harbor aging conference, the festering problems exploded. Kaeberlein spoke in the first session. Resveratrol did not key caloric restriction, he said. Resveratrol did not stimulate Sir2 to act on any of its normal targets. It only appeared to activate the sirtuin enzyme due to the mechanism of the company's fluorescent kit. It was a "substrate specific activator." The only way to prove the compound's significance, Kaeberlein said, was for resveratrol to extend life in mammals. Sinclair angrily questioned their methodology, and the question-and-answer session grew heated. "I'm on to mammals anyway," he said later.

For solace, David Sinclair reread the historian of science Thomas Kuhn's classic, *The Structure of Scientific Revolutions*. Kuhn suggested that normal science tends to be highly conservative, nibbling at the edges of a dominant theory, and that change happens only with rebellion, much like a political revolution. Science change begins with a series of anomalies a theory cannot explain, building stress until the social organization of a field can no longer reject new anomalies out of hand. "I actually went through Kuhn's list of traits that identify a paradigm shift and I'm going, yep, well we've got rebellion, yep, we've got chaos," Sinclair said. They were living, it seemed, through a real, unfolding, new paradigm in the history of science, with exactly the battles over terminology, assumptions and method that Kuhn predicted.

The science leapt to humans, as several teams raced to track the insulin gene in human populations of centenarians. "Maybe there's a synthesis emerging that will in five years become dogma," Sinclair said at the end of the year. No one was banking on it.

7 *Betting the Trifecta: 2005–2006*

By 2005, fifteen or so major longevity genes had emerged as laboratories ranging from Rutgers to Northwestern to the huge new Barshop Institute in Austin, Texas, to the Wisconsin Regional Primate Center in Madison sought to raise money for more research. From London to Munich to Athens, more institutes raced to stake their own claims. The genetics of longevity, while still a young field, was bidding to become a leading edge of molecular biology. Some of its academic conferences shifted from obscure gatherings to slick conclaves at huge resorts. The century-old Dutch scientific publishing company, Elsevier, for instance, sponsored genetics of aging retreats in Europe's classical cities, like the one on Crete, underwritten by pharmaceutical companies like Novartis and GlaxoSmithKline. In so doing, the companies sought an early glimpse of the most marketable discoveries.

As more labs sought to understand the exact processes of late-life fitness at the molecular level, the picture became more complicated, even as the theories of aging—of free radicals, limited calories, inflammation, and DNA repair—promised to simplify and converge. The lab experiments "made the disparate theories of aging fuse together," Steven Austad commented, somewhat hopefully, into a focus on antagonistic pleiotropy, the idea that what is good for us in youth is bad for us as we age.

A lifetime could be divided into three phases: growth, healthy adulthood, and decline. Gene networks increase in strength and "fidelity during growth," said Wayne State University's Robert Arking, in "maintenance during adult health, and then experience progressive loss of feedback during senescence, destabilizing cell functions as the forces of natural selection wane." Many molecular mechanisms either preserve or shorten life. Some inhibit repair mechanisms, some assist. Some kill diseased cells, others killed off healthy cells. Some cells do not turn over proteins at the youthful rate. Some hormones rage. As theorists of aging caught up with lab discoveries, scientists, led by a glimpse of a possible paradigm, adapted new instruments to understand the ways in which the discoveries might give insight into health.

Numerous discoveries suggested that the insulin pathway was the key and that tweaking it could improve the quality of aging in humans. At Rutgers University, biologists Laura Herndon and Monica Driscoll investigated the ways in which insulin-sensitive worms resisted sarcopenia, the decay of muscles, as they aged. "They look astonishingly human!" Driscoll told an interviewer. "They experience midlife onset, progressive loss of function, and decline in strength." Driscoll began testing two interventions to improve muscle strength. One was the diabetes drug metformin, under investigation by Elixir, and the other was a collaborative effort with the nonprofit International Biodiversity Group to test extracts from plants all over the world. At Northwestern University, biologist Rick Morimoto discovered that the insulin mutants resisted Huntington's disease. At the University of California, San Francisco, Kenyon's lab showed that her mutation in *daf-2* inhibited the growth of precancerous tumors. At a Boston University lab, flies engineered to lack the insulin pathway gene did not get heart disease. (Fruit flies, *drosophila*, are subject to the same decay of arteries as humans.) At Southern Illinois, Andrzéj Bartke showed the insulin pathway worked separately from caloric restriction's longevity effect in mice. When combined, they doubled life span. For this, he won the coveted Methuselah Mouse Prize from the Methuselah Mouse Foundation. Taken together, the discoveries suggested ways in which overlapping gene pathways could make late life better than we had imagined, at least in the lab.

daf and *SIR* became "celebrity genes," according to Steven Austad. Controllers of many life-lengthening genes or hormones, they tempted everyone to choose one or the other and try manipulating them with a compound. "We were almost like a religious cult with these two pathways," said Ruvkun of the debates, praising the National Library of Medicine for its high-tech free website to compare immediately newly discovered longevity genes with any known gene sequence in any species. A new field called the computational biology of aging studied the various signals as information moving through a system at the level of molecule, cell and tissue. "The problem was integrative," said Marc

Tatar. "Like Venn diagrams, you had to find what was common in aging between different organisms."

Aging and life span, the oldest of biological questions, caught fire. "The genome was built to evolve quickly," Kenyon said. As she told the William Harvey Group in New York, a nonprofit named for the eighteenth century discoverer of the circulation of blood, which awarded her its prestigious annual award in 2005, "The path to longevity is within us in the forms of genes and networks, waiting to be nudged in the right way."

Four Pathways

By 2005, a host of third-generation researchers, along with those inspired by the new science, competed to discover more longevity genes. The new candidate genes included odd names like target of rapamycin (TOR). Of these, four intertwined genetic pathways came to the forefront: insulin, sirtuins, the anticancer gene TOR, and the energy regulator AMP-kinase. The key was to find which worked best in humans.

The human version of *daf-16*, *FOXO*, made a favorite new target, and Anne Brunet, a young woman at Stanford whom Kenyon inspired, was the expert *FOXO* hunter, studying the different outside stimuli that extended life. In her postdoc at Harvard's Michael Greenberg lab, Brunet laid the groundwork on mammalian *FOXO*. Later in her own lab, she discovered that caloric restriction started in middle age extends life by activating AMP-kinase, which in turn activated the FOXO proteins. These proteins were already being studied as cancer treatments. Those proteins proved crucial to cellular processes regulating programmed cell death, stress resistance, and protection against oxidative stress.

Brunet was one of a new wave of researcher, younger yet high-powered and committed to the study of single-gene longevity interventions. A native of France, she received her undergraduate degree at Ecole Normale Superieure Paris in molecular biology, and her PhD in cell biology at the University of Nice. In 2004 she became an assistant professor of genetics at Stanford University's School of Medicine. Funded by the Alfred P. Sloan Foundation, American Foundation for Aging Research, American Institute for Cancer Research, Crain Tumor Society Foundation, and the Esther A. & Joseph Klingenstein Fund (focused on epilepsy), the Brunet lab demonstrated the new public and private money driving the molecular study of longevity.

At the University of Fribour in Switzerland, researchers found that decreasing the activity of a new gene called target of rapamycin (TOR) more than doubled the life span of worms. At the University of Washington, Matt Kaeberlein and Brian Kennedy were screening gene deletions and also identified TOR. The drug rapamycin, discovered on Easter Island and used to suppress the immune system in kidney transplant patients, decreased the activity of the gene. Decreasing TOR dramatically was eventually shown to increased life span in worms, yeast, and mice. TOR controlled growth and the rate of protein synthesis and suddenly became a leading longevity candidate that would soon take center stage.

At the University of Wisconsin, the Weindruch lab had kept some forty-eight rhesus monkeys on a low-calorie diet for twenty years. To market their discoveries, researcher Rick Weindruch and Tomas Prolla cofounded a company called LifeGen. They invested in and secured the rights to use GeneChip, the plastic chips that resembled computer microchips. They could at a glance tell you how many of the twenty-one to twenty-two thousand or so human genes were turned on in a cell. With the GeneChip, the researchers conducted a huge microarray analysis showing that some interventions make animals live longer without having an impact on the basic core of the aging process. One early surprise showed that only a small number of genes actually changed in their activity as the animals aged. This result contradicted evolutionary theories of aging, which argued that larger numbers of genes must be involved in the processes of decline. Perhaps one need only nudge one or two gene pathways to lead to longer, healthier lives.

That idea of "nudging" gene pathways got more credit from fly researcher Linda Partridge at the University of London, who was studying the interaction between FOXO and longevity, which she claimed "tipped the balance towards survival."

Still, the National Institutes of Health still would not fund longevity gene research in significant amounts. "What is needed is an Apollo-type government investment in biology of aging research," said founding NIA director Robert Butler. The big pharmaceutical companies remained skeptical, as did many university researchers who specialized in various aging diseases. As molecular genetics conferences became more competitive, one saw the same qualities of the molecules as in the people studying the molecules, each fighting to prevail.

More money emerged from an unexpected source, though, as cosmetics companies latched onto the new longevity gene science. Avon developed a skin cream based on the human version of Sir2. Estée Lauder provided funding for the Sinclair lab and sponsored research by academics like Guarente, who was increasingly dissatisfied with Elixir for not following up on sirtuins and was on the verge of leaving his own company.

The Rutgers worm lab of Monica Driscoll, for instance, adapted a Johnson & Johnson company technique for evaluating its skin-care products to a new lab technique for measuring aging by gauging the rise in lipofuscin levels, yellowish granules linked to dementias and macular degeneration.

In 2006, Avon expanded its line of skin creams based on sirtuins. Estée Lauder introduced an antiwrinkle moisturizer based on the SIR gene. "The Time Zone Line and Wrinkle Reducing Moisturizers promises to be one of the first of many products from Estée Lauder to feature sirtuins," said Daniel Maes, Estée Lauder senior vice president of Research and Development, who called sirtuins the "next big thing" in antiaging treatments. The company claimed that sirtuins battled the visible signs of aging by "slowing down the rate of cellular division in the skin to give cells the time to perform necessary repairs." The market was potentially huge. Even Harvard, which had famously failed to cash in on its professors' discoveries, retooled its technology transfer office to better protect its patent interests. "This is the moment," Sinclair told the *Wall Street Journal*, as investors put up another $53 million in a second round of funding for his company, "when life span becomes the race to split the atom."

Backlash

With the early hype came a backlash. "Aging is not programmed, period!" declared the diminutive Steven Austad, in his wild rhinoceros tie, black suit, and jogging shoes, to reporters at the February 2006 meeting of the American Association for the Advancement of Science in St. Louis.

Biotechnology companies were overselling their claims, claimed Marc Tatar. "How can you contradict your hypothesis if you've got $50 million of someone else's money riding on it? You can contradict it but you're sure not going to tell anyone." Resveratrol did not stimulate Sir2's activity in most yeast strains. Polyphenols were known to give false-positives. Leading the criticism were Matt Kaeberlein and Brian Kennedy. Their finding that in three different yeast strains resveratrol caused no life extension appeared in the April 2005 issue of the *Journal of Biological Chemistry*. "The mechanism accounting for the putative longevity effects of resveratrol should be re-examined," they concluded.

One problem, it was argued, might be dosage. "Resveratrol is very tricky, very temperature sensitive," David Sinclair responded. "I can't help it if they can't get it to work." His lab, with Marc Tatar and Stephen Helfand, had extended fly life span some 20–25 percent with a lower dose of resveratrol. For the moment, the criticism stopped.

Linda Partridge and David Gems, however, tried to repeat the resveratrol results in flies and could not. There was a growing concern about publication bias toward positive results. If an experiment showed a negative result, no one was interested. Others went further in criticizing the influx of money into molecular genetics. By 2005, the editor of the British medical journal *Lancet* accused top science journals as having "devolved into information laundering operations for the pharmaceutical industry."

Criticism also cropped up of the claims that the sirtuin genes were the key to caloric restriction. The Sir2 protein was certainly a NAD-dependent deacetylase, but the question was, did NAD signal precisely the presence or absence of food. Some of the experiments were unclear. The experiments in yeast showing that caloric restriction activated sirtuins were flawed, because the marker used was regulated by glucose independently of sirtuins. Ultimately, it appeared some of the effect depended on levels. If NAD levels went up a lot, the yeast did not live longer. If it went up a little, they did.

On a larger level, several critics argued that the whole idea was oversimplified or wrong and that life-lengthening interventions worked only on lab animals, not on real wild animals. The long-lived strains were, in a word, wimps. Some were sickly or sterile. When Buck Institute researcher Gordon Lithgow mixed in *age-1* and *daf-2* mutants with normal worms and starved them, the normal worms survived, where the longer-lived ones did not. But when he raised the temperature to 90° Fahrenheit, the mutants outlived the normal worms. Steven Austad and Rich Miller began doing life-span experiments on wild mice. The reality, critics countered, is that many competing genes, switches, and biochemical processes affect aging.

As the economy faltered, Elixir worked to antagonize the hunger hormone ghrelin. Inhibiting the cell's ghrelin receptor improved insulin sensitivity, cholesterol levels, and resistance to obesity. The company struggled for money. Cynthia Kenyon and Lenny Guarente gave up much of their equity share as their company fought for more funding. The investment firm MPM Capital became the biggest shareholder and took control of the board. The company had run through a lot of capital. It had interesting compounds. It had a sirtuin inhibitor ready for trials in treating Huntington's disease. It had an antagonist of ghrelin, a compound proven in three trials to extend healthful life span of mice fed a high-fat diet. Such a compound could be marketed to combat obesity and metabolic diseases like diabetes. But Merck and Pfizer had chased that goal and failed. Elixir was facing long, expensive preclinical drug development.

At Sirtris, researchers were also worried. Though adept at raising capital from well-heeled investors via a combination of bravado, altruism, and, paradoxically, reminders that investors could lose all their money, the young company still could not point to a single verified successful life-span intervention. They had little life extension in mammals. They needed better assay results in cells. Then they needed results in animals, preferably for specific diseases. The company said little in public and was very deliberate in hiring. It moved into a cramped, nondescript building in Cambridge, Massachusetts.

Then came another gene, or rather, an old gene reappeared—one that changed the course of the idea.

Easter Island

On a windswept beach of Easter Island in 1967, beneath the giant carved wooden heads that for centuries had stared out to sea, a young field scientist scooped a handful of dirt among thousands of other samples. Researchers from the Canadian Medical Expedition was digging through samples of soil and rocks, seeking new antibiotics in the battle against resistant bacteria. Taking the samples back to Montreal, they found a natural compound that was eventually named "rapamycin" from the native name for Easter Island, Rapa Nui.

Rapamycin proved to be an antifungal compound that GlaxoSmithKline, Novartis, and others studied as a possible anticancer or antitumor drug. But more than that, it was found to suppress the human immune system and could be prescribed for transplant patients. It almost never made it to market because the company that developed it, Ayerst, went bankrupt. The head researcher had the foresight to harvest a final batch from the fermentation tanks and store it safely when he was transferred to what became Wyeth's Princeton laboratories. Steady progress was then made, and rapamycin was licensed by the FDA in 1999 for use as an immunosuppressant drug in kidney transplant patients. Highly prized, rapamycin inhibited the body from reacting to the new, foreign organ after the transplant of a new kidney.

The University of Basel, Switzerland, researcher Michael Hall had originally discovered the gene affected by rapamycin in yeast, and named it target of rapamycin. No one expected that eighteen years later, the same gene would come up in three longevity experiments: one in worms, one in flies, and one in yeast by Matt Kaeberlein and Brian Kennedy. TOR turned out to be an essential component of the nutrient-sensing pathway, coordinating growth and protection against cancer. The most significant life-span result came in mice, where the gene to be called mammalian target of rapamycin (mTOR). mTOR proved to be a longevity gene that intersected and overlapped with the insulin pathway.

In Bandera, Texas, the "Cowboy Capital of the World," a group of researchers had developed the NIA Interventions Testing program, which took the newest antiaging interventions and tested them in three different mice labs using the same strains for uniformity. In 2004, the University of Texas researcher Dave Sharp proposed rapamycin be one of the third set of compounds tested. His colleague Randy Strong designed the experiment. But the compound was highly expensive, difficult to manufacture, and even more difficult to administer in the proper dosage. It kept breaking down chemically in the mouse food pellets.

After nearly two years, Strong finally turned to a San Antonio company to encapsulate the rapamycin. By that time, however, the mice were already twenty months old, the equivalent of a human at sixty years. Yet those twenty-month-old mice, treated with rapamycin, lengthened their average life expectancy more than 20 percent. The drug also extended the life span of flies. It was like a sixty-year-old human taking a drug and getting twenty years of health beyond the usual ten. Rapamycin was an available drug, already tested in humans. The gene it affected turned out to code for a major amino acid and nutrient sensor that blocked salvage pathways when active. mTOR was a major nutrient sensor in animals. It made a protein that became active if you eat. If you inactivate the protein, which the drug rapamycin did, partially through the insulin pathway, you could multiply ribosomes (cell machines that make proteins), increase the stability of amino acids, trigger autophagy (cellular recycling that happens in caloric restriction), increase protein synthesis, and inhibit the translation of proteins associated with obesity.

They already knew rapamycin lengthened the life spans of worms and yeast and that it worked through the caloric restriction pathway, with some overlap with the insulin pathway. At UCLA, Valter Longo was inhibiting TOR in yeast and seeing that, in combination with calorie restriction, he could extend yeast life span tenfold. Attention turned to its possible efficacy as a longevity drug in humans. The problem was that it cost about $1,000 dollars for a week's dosage.

In labs around the world, researchers sought to understand the action of the human version of the insulin gene target. As in the worm, in humans the gene made a transcription factor that controls other genes involved in longevity. "It's like a superintendent of a building," said Kenyon, "hiring the mechanics, painters, carpenters, plumbers, janitors, and maids that keep it young." At Albert Einstein Medical Center, researchers were finding their Ashkenazi Jewish population shared a *FOXO* gene variation similar to that of the worm's long-life transcription factor. Studies had begun in German, Japanese, and Scandinavian populations, with several pointing toward a similar variant of the regulating gene in the FOXO pathway, as well as other genes. If insulin made a hormone that causes wear and tear, it made sense that an intervention turning down the use of insulin might extend healthful life. "It is almost certain that various means of slowing or delaying aging . . . will be eventually tested on humans," concluded Robert Arking, author of the basic field textbook *The Biology of Aging: Observations and Principles*.

Powerhouses

In late 2006, a Harvard researcher shifted attention to the most ancient repository of DNA, the cell powerhouses or energy factories called mitochondria, and to the most accepted molecular mechanism of aging, the free radicals they create. Mitochondria held a special place in the effort to understand genes and their control of aging. Once free-standing, anaerobic, one-celled organisms, these sausage-shaped organelles became incorporated early in evolution into the modern cell in a primordial corporate merger. In doing so they retained a small number of their own genes, separate from nuclear DNA, passed down directly from mother to children, one generation to the next. Mitochondrial DNA made possible the great search for Eve, mother of all humans, and helped scientists to trace her to East Africa. Mitochondrial DNA also mutates more than nuclear DNA, presumably because it lacks the repair mechanisms of the cell's main control center. Because of that high mutation rate, and because of the mitochondria's generation of the power that runs the body and the free radicals these powerhouses produce, they were the main suspects for the damages of aging. "We are the accumulation of our mitochondria," wrote the biologist Lewis Thomas.

A study of resveratrol's effect on the cell's energy factories was hitting stride. At Harvard, Pere Puigserver, a young, rumpled, thoughtful researcher, used mice to study the ways in which nutrients and hormone signals regulate different metabolic pathways. Since mitochondria are the cell's energy factories, they are the engines of metabolism, problems in which could lead to diabetes, cancer, and premature aging. Puigserver's team focused on the way cells sensed changes in nutrient and hormone signals, prompting them to make transcriptional, or DNA-copying, changes.

Puigserver showed that treating human cells and mice with resveratrol increased the number and health of mitochondria in aging cells. The key to resveratrol's effect involved the protein complex PGC-1αfl a master regulator of mitochondrial metabolism. The discovery suggested that resveratrol triggered a coordination of signals through SIRT1 and PGC-1αflAging was a process of losing metabolic efficiency. PGC-1fi increased the efficiency of mitochondria. It increased the size of mitochondria in humans. The effects were reminiscent of caloric restriction, with its highly coordinated symphony turning on certain health-extending genes. Puigserver's discovery offered a mechanism for how their resveratrol-inspired drugs actually might work.

When the discovery appeared in *Cell* in December 2006, spirits picked up at Sirtris. "That was what made me decide to devote my life to this," said the company's new director of science, Michelle Dipp. "It was not just one intervention. There were thirteen places" on the molecule where SIRT1 worked. "We could now address a good simple disease." The disorder was a little-known disease called MELAS syndrome, a progressive neurodegenerative disorder caused by a mutation in mitochondrial DNA. Now Sirtris could apply for FDA approval to begin clinical trials for a resveratrol mimetic.

Sirtris scientists had published a more popular paper in *Nature* the previous month, showing that resveratrol, the SIRT1 activator, could reduce the impact of a high-fat diet and increase stamina twofold. A video of fat, running mice on a treadmill was a popular hit on YouTube and a cover story in the *Wall Street Journal*. In it, the mouse on resveratrol ran almost a kilometer further than his comrade. The obese mice showed "amazingly youthful vigor as they aged," wrote the *Journal* reporter David Stipp, who detailed the dispute as to whether the compound worked in the same molecular pathway as caloric restriction. The discovery put Sinclair on the *Charlie Rose* show and into the *New York Times*.

On the strength of those findings, the company kicked off a second round of financing in early 2007. Boston Red Sox owner John Henry and Merrill Lynch's cofounder, Peter Lynch, invested part of their private fortunes in Sirtris. Westphal planned to take the company public in a few months, an extremely fast turnaround.

Their main challenge remained to find a patentable, synthetic, more precise and powerful version of resveratrol. Sirtris estimated that a person would need to drink a thousand bottles of red wine a day to obtain an effect. To address that problem, a twenty-five-person team was testing much smaller molecules with much higher rates of efficacy than the plant extract. The company had already developed a purified form of resveratrol they called SRT501, which they claimed was seven times more powerful than the natural form. The discovery of PGC-1α's role, the development of SRT501, and most of all the video of the running obese mice, opened the market more. "The whole field," said Dipp, "went gangbusters."

At the same time, Elixir sought to design drugs that could act as either triggers or blockers of the ghrelin receptor, pushing to get results in higher mammals. It had promising results in obese mice. The researchers grew more excited. "I'm thrilled," said Ana Maria Cuervo at Albert Einstein Medical Center. "It's a whole new science of first principles," said Brian Kennedy. "[It's] a revolution in our capacity to interfere [with aging]," said Monica Driscoll, who compared the worms to the Whos in Dr. Seuss's *Horton Hears a Who*. You could hit several longevity pathways. Andy Dillin and Suzanne Wolff speculated about betting a "trifecta of aging" in a 2006 review article. But what happened when you took antisocial, difficult, brilliant scientists out of their labs, threw a ton of money at them, and then expected them to perform antiaging miracles?

PART 2: DEFYING GRAVITY: THE BATTLE TO SELL THE IDEA OF EXTENDING LIFE

In a remote Trinidadian mountain stream, forty-two-year-old David Reznick wondered why he was spending years studying guppy life span. But what the University of California Riverside researcher found was startling: wild animals' life spans could lengthen or shorten within a short time. The rate of aging in the wild slowed or sped up as required, relatively quickly. It could be inherited. Chapter 8 covers the research in the plasticity of aging and the race to apply those discoveries to human populations.

The scientific and business battle to sell the idea of extending life reached a crescendo in the years 2007–2011. Chapter 9 examines the unprecedented media frenzy. *Oprah* and *60 Minutes* became the soundstages of science. The show, and not the conference, became the place where research was performed and communicated. In 2007, bankers surrounded Sirtris's Christoph Westphal as the company prepared to go public. In 2008, GlaxoSmithKline bought Sirtris for $720 million. The only problem was that the original science was flawed. The result threatened disaster for some and fortune for others.

Chapter 10 explains the growing number of disease therapies coming from the insulin pathway, focusing on the work by Judy Campisi on cancer at the Buck Institute on Aging and by Andy Dillin on Alzheimer's at the Salk Institute. The explosive reports from Amgen and Pfizer scientists that some sirtuin results could not be repeated were reported, as federal funds slowed. Other private sources of research money emerged, such as Estée Lauder, the Paul Glenn Foundation for Medical Research, and the Larry Ellison Medical Foundation. PayPal founder Peter Thiel funded the Methuselah Mouse Prize, and Dole Company magnate David Murdock financed a billion-dollar research institute in North Carolina to understand the impact of fresh food on healthful aging. The effect of the some of the SIR fallout paradoxically helped make the science become white-hot.

Chapter 11 returns to the best of the research on insulin, the gene mTOR, *SIR*, and the enzyme AMP-kinase as a new field focused on gene switches, called epigenetics, in the fastest growing demographic group in the world, centenarians. Chapter 12 limns the effects of the original idea as gerontologists sought to develop a new theory of aging in spite of lab science's setbacks, confusion, and complexity. The new world of longevity genetics tools developed at the beginning of this decade ratcheted a young science into a new prominence. The previous twenty-five years had set the stage. This chapter offers a scorecard of the most significant emerging discoveries. Chapter 13 reviews the lessons of the medical science race.

8 *Sex, Power, and the Wild: 2001–2008*

On a hot afternoon on Trinidad's Yarra River, David Reznick was carefully trying to carry a bag of guppies without losing the contents. In a Panama hat and torn T-shirt, the University of California at Riverside teacher swatted at mosquitoes and shielded his eyes from the sun. For four years he had been studying the effect of environment on guppy life span. But what Reznick found was startling: wild animals' average life span could lengthen or shorten significantly in a short time, and that lengthened span could be inherited within a couple of generations.

Far from the money and ego of the genetics laboratories, researchers in the wild discovered that animals lived longer in more different ways than we had ever imagined. Some animals mastered the environment to live longer, as evolutionary theory predicted. Some grew wings, like bats; some grew bigger, like tortoises and elephants, some got smarter, like some humans or other primates; and some lived on safe islands like opossums or mountain tops like Joshua trees. Some grew smaller and burrowed underground. Some combined traits, like parrots, which are both intelligent and winged and live for forty to seventy years. Some developed social skills or physical strength. Some simply lasted, like the thousand-year-old California sequoia or Norway spruce. Longevity in the wild, it turned out, happened all the time.

The effect of this research was exhilarating. Previously, evolutionists thought safety was somehow required for longevity. But Reznick showed the opposite: the longest-lived wild guppies came from the most hostile environments. By comparing the life spans of fish in safe mountain backwaters with those in predator-heavy pools, he saw that danger appeared to improve fitness and reduce competition for food. He even showed that older female guppies experienced menopause and passed on their longevity to their children. Even female gorillas, it was seen, experienced menopause. "Their pattern of growing old is remarkably similar to that of mammals" when removed from their high-predation environment, Reznick said of his Caribbean guppies. His 2005 discovery got picked up by MSNBC. "There's more to life than making babies!" read the banner headline on its website.

Perhaps the most amazing thing about the discoveries of longevity in the wild was that they had not happened sooner. One reason we know so little about natural aging was that such research made for grueling, painstaking study in the world's remotest places. It was difficult to see natural longevity because it happened over many years in jungle rivers, deep-sea vents, skies, African savannas, and tropical bayous. The sea, for instance, made a rich breeding ground for old animals. Patagonian toothfish (or Chilean sea bass, as you may have seen it on your restaurant menu), reach sexual maturity after forty and live longer than humans. "Don't order it," the noted California researcher Leonard Hayflick told me. "It could be one hundred years old." Some rougheye rockfish live two centuries, as do some bowhead whales. Sea birds like albatrosses and storm petrels lived up to one hundred years in remote stretches of ocean. Weird giant worms at deepwater cold seeps in the Gulf of Mexico live two hundred and fifty years. Alligators in Florida's Everglades and turtles on the Indian Ocean's Seychelles Islands grow and reproduce until they die at ripe old ages.

Life span in nature made a shifting and dramatic quality that could be passed down, which means that genetic mechanisms had to drive it. The question was which ones.

The idea seemed tantalizing, except for one big problem. Evolutionary biologists like Reznick vehemently rejected the notion that tweaking a single gene or gene pathway could affect life span. "The genome is not a symphony, but rather a family of chemicals vying with each other, with many conductors and many orchestras," Reznick told me. Even something as universal in mammals as the placenta, he once wrote, could have evolved independently at different times around the world. If that was true, then something as vague as longevity certainly was not a unified process governed by shared genes. "The lab researchers hate this," repeated Steven Austad. "Aging is not programmed, period!"

For years these two opposing groups—researchers in the wild and those in the lab—did not speak. Then they did. What happened next altered an idea's prospects. To understand how, first one had to look back.

Hopeful Monsters

For most of history, life in the wild was assumed to be nasty, brutish, and short. Old age and infirmity was assumed to end with swift, often violent, death. Since few animals reached old age, few researchers considered what it might look like.

To say that wild animals evolved longer life spans raised the question of how they did so. Is adaptive life-span change driven by single shared genes or by enormously huge networks of almost random events? Did one longevity gene spread its influence around the world as, say, the developmental genes that build the backbone, or did longevity evolve in different ways in different places? If animals evolve longer life naturally, the mechanistic questions seemed so highly complex as to be impossible to test. Still, in itself, the idea that single gene changes created wholesale new life forms or life qualities was not new. It had been offered in 1940 by German geneticist Richard Goldschmidt and was quickly ridiculed as a belief in "hopeful monsters."

In the late 1990s, Ruvkun, Kenyon, and others cited as their support the developmental revolution, when it was seen that the similar clusters of genes controlled thousands of processes to direct the growth of many different animals. Ruvkun described the Nobel-winning HOX gene discovery in the fruit fly as "luminous" and exactly "what we were trying to emulate." In the 1980s, it was seen that seven bodybuilding genes align in a fruit fly embryo's cell the same exact way the body parts they built lined up in an adult fly. Then researchers saw those same seven genes direct growth in exactly the same choreography in animals ranging from worms, as Kenyon discovered, to mammals, including humans. The genes did so by sharing an identical stretch of DNA, which produced a protein that turned on and off other genes in the same sequence. The mechanism was widespread, common, and beautiful. We were all, in our way, hopeful monsters.

In fact, when researchers revealed that another single gene, *PAX6*, created the vastly different kinds of eyes of flies and owls and nearly blind opossums and humans, the pioneering biologist Sydney Brenner wrote a humorous story portraying Francis Crick accosting God in heaven. In the essay, God is an absent-minded, paint-spattered maintenance worker in overalls. When asked how he invented the fly's multifaceted eye, he replies, "Well . . . um, actually we don't know but I can tell you we've been building flies for 200 million years and we have had no complaints."

Early on, when Kenyon and others cited the discovery of a shared developmental program in flies, worms, mice, and humans as a model of a potential shared aging program, the critics pounced. She was mistaking development, planned and uniform, for aging, which is disintegration as random as the wrinkles on a face. Development was "anabolic," building complex molecules from simpler ones, said Leonard Hayflick. Aging was catabolic, breaking complex molecules down. How could breakdowns evolve?

The argument returned to the relatively straightforward definition problem between longevity and aging. Aging could not be regulated, many said, though it could be perhaps patterned a bit by genes or anything else. But longevity or healthful life span could be regulated—in fact, it had to be. If times are harsh, organisms hunker down, Kenyon said, living longer until conditions improve enough for them to reproduce.

Others pointed out that tiny worms may have a gene program to extend life, but they only travel a couple feet in a lifetime and use simple genetic programs to hibernate and reproduce and everything else. For humans and other mammals with trillions of cells, each with tens of thousands of proteins fighting stress, repairing damage, and preserving health, life span could never be altered by tweaking single genes. Mammalian gene regulation was far more complex than gene sequencers could detect. One gene can have many effects, in different tissues, at different times, under different conditions. If human longevity was timed by genes, why do our life spans vary so much?

In a growing backlash against the post-genome determinism in popular discussion, the University of California medical researchers Linda and Edward McCabe gave an example of just how difficult it was to apply a single gene discovery to the problem of human health. The couple studied a rare, simple, lethal disease called glycerol kinase deficiency, which affects infant boys. Since the disease was produced by only one gene, they thought they could analyze the gene to devise a test that would give parents a chance to save a child. The McCabes assumed that the most serious form of the disease would show the most serious gene mutations. In fact, the couple found no difference in health by the degree of mutation. They wrote, "some of the healthy individuals actually had bigger mutations." The effects of an individual's genes, they concluded, are shaped by experience, diet, environment and, even, sheer chance.

Given the limits of laboratory research, the best way to analyze the mechanisms of longevity in the wild would be to study them in the wild. "We need to look at the long-lived animals that evolution has produced," said Steve Austad, "and see what made them successful." This effort made harrowing work, a "herculean slog," commented Marc Tatar, referring to David Reznick's research. "I'm finished with it," Reznick told me of his life-span studies, remarrying and raising two young children. Named to the American Academy of Arts and Sciences, he expanded his experimental evolution project in Trinidad. Of the study of natural longevity, he said, "I don't have time in my life anymore." But others did.

Aging in the Wild

The study of aging in the wild featured a colorful cast of disaffected adventurers. In the 1960s, the mathematician and Gerontological Society of America president George Sacher filled his Berwyn, Illinois, ranch home with long-lived squirrels, bats, muskrats and other animals in order to study the biology determining their life spans. In the 1980s, *Newsweek* reporter Cynthia Moss became so captivated by long-lived Kenyan elephants (seventy-five years or more) that she quit her job to devote her life to the animals, founding the Amboseli Trust for Elephants and writing a memoir. Moss observed that older males reproduce only in their forties and fifties, and only when they earned status gained by time serving the group. Elderly females taught effective response to threats like lions. These and other studies caused researchers to speculate on the adaptive advantage of old age and menopause, living past our ability to reproduce.

The popular "grandmother theory" suggested the survival of early human settlements got a boost from older women giving up reproduction to help children raise their grandchildren. One study found that on Simeulue Island, west of Sumatra, households with older family members recognized earliest the warnings of the 2004 Indian Ocean tsunami and moved quickest to safety in the highlands. University of California demographer Ronald Lee devised algorithms to show the benefits of this longevity dividend. A new field called kinship dynamics sought hard science to bolster the idea.

However, save for some suggestive demographic data on survival rates in history, relatively little hard evidence could be found in archaeological records that the grandmother theory was true. The study of long-lived wild animals, by contrast, suggested that lab research could understand the gene and protein changes that lead to lengthened life naturally. Richard Miller and Steven Austad did just that by studying wild mice whose cellular biology, it turned out, looked much like that of the less extreme, long-lived lab mutants. Wild mice were smaller, leaner, and more insulin-sensitive than lab mice, sharing many of the same growth factor defects as the long-lived lab mutants. They lived longer than lab mice.

Their secret seemed to be stress resistance. The cells of wild mice resisted damage due to illness, heat stress, radiation, free radicals, or oxidants. The mechanism appeared to be epigenetic, meaning it was a thermostat set in infancy by the low insulin and growth factor circulating in their bodies and passed to the next generation. The two other long-lived lab mutant mice, the Ames dwarf, which shared the qualities of small size and insulin sensitivity, and the growth hormone receptor knockout mouse, also shared the same cell-stress resistance. Resistance developed in the first few months and was accompanied by high insulin sensitivity. Cells from calorie-restricted wild mice did not show a similar resistance, suggesting the life-span extension induced by low-calorie diets differed from that of the dwarf mice. Because the long-lived animals shared similar low insulin and growth hormonal patterns, longevity and cell-stress resistance, the long-life pathways may have evolved early in the history of life. Maybe mammals had a longevity window of opportunity in infancy.

Sea birds had a similar resistance to cell stress, but their mechanism appeared to be related to telomere length. The albatross and Leach's storm petrel in the extreme north and south Atlantic seemed to maintain their chromosome endings as long as they lived. The petrel, in fact, made for one of the only known animals whose cells' telomeres lengthened as they aged. Many long-lived sea gulls showed no signs of aging and remained free from cancer. A haunting set of photographs shows the same long-lived gull studied by the Scottish ornithologist George Dunnet in his twenties and again in his sixties. As the man ages from youth to white-haired, wrinkled senior citizen, the Orkney Island gull looks quizzically back at him, unchangingly youthful in physique.

The third great insight into natural longevity came from the east African savannas, where the personable naked mole rat remained free from cancer even at the end of their spectacularly long lives. Living in complex, three hundred-animal societies in cool, dark tunnels seven feet below the arid grasslands, they had the longest average life span of any rodent, about thirty years, five times longer than expected based on their size. The equivalent of one-hundred-year-old humans, the tiny animals keep reproducing until the end of their lives. Male and female adults cared for the pups. University of Texas researcher Rochelle Buffenstein found their cells resisted a broad array of harsh toxins. They had low levels of insulin and high insulin sensitivity, much like long-lived bats. But amazingly, the animals also had high levels of oxidative stress and short telomeres. What naked mole rats had was a stress-protective version of a common gene called Nrf2, which preserved the power of proteins to fold properly. Many late-life illnesses are diseases of protein folding. As we age, the tens of thousands of proteins in our cells struggle to maintain proper shape. Heat, chemicals, and other stresses attack and degrade proteins, something like frying an egg. It is the job of some chaperone proteins to unfold and refold their aging counterparts.

Taken together, the discoveries suggested that several paths could lead to healthful old age in the wild. Longevity could happen in both high-safety and high-predation environments. It could be inherited, so something was being passed down by genes. But "just because the same insulin receptor gene lengthens life in flies, mice, yeast, worms, monkeys, and humans, does not mean it has a shared origin," said Reznick. "Finding a single gene pathway related to a single characteristic is simply finding one component in a complex system."

The question remained open. But the vehemence of the argument demonstrated the "the power of the idea," observed Kenyon, whose lab showed that the FOXO transcription factor controlled hundreds of genes for resistance to oxidative stress and infection, freedom from the cancer-type tumors worms develop, and what one might call muscle disease (sarcopenia) and dementia in a worm. "It is truly a physiological shift, a hunker-down state," she said. Nutrient or environmental sensors triggered a long-life setting. Northwestern worm biologist Rick Morimoto seconded the notion. "I absolutely believe that aging is programmed," he told me. "It's not chaotic, it's not random. It's just very complicated and intricate," concluded Kenyon graduate Joy

Alcedo, now at Wayne State University in Detroit. "It all depends on whether we understand it."

Perhaps aging is not programmed, but healthful longevity certainly might be. Genes influence the rate of aging by slowing it, molding it, or postponing it. The key was to find the same pathways in humans. A clue came from the Second World War.

The Human Clock

In the long, frozen winter of 1944, the Nazis, furious over a Dutch rail strike in support of the Allied invasion at Normandy, retaliated by flooding the country's great industrial region, including Amsterdam and Rotterdam. As the Allied army got bogged down, food supplies in Holland dwindled to nothing. The Dutch people starved, eating rats, mice, and tube roots. Never had an industrialized society broken down so completely.

In the years following the war's end, researchers noticed that children conceived during Holland's great *Hongerwinter* (Hunger Winter) shared similar health problems in later life, including diabetes, obesity, and heart disease, all at an earlier age than their counterparts. The actress Audrey Hepburn attributed her clinical depression to being an infant during the famine. The data seemed to support a theory of the University of Southampton's David Barker: changes in life's thermostat in the womb, created by molecular gene switches, may account for much of our health or sickness as we age. In Leiden, Holland, a young doctoral candidate with a strong sense of life's humor, Eline Slagboom, became fascinated by the mystery. "It just seemed to be a ticking time bomb," she said. Slagboom set to find out what actually happened in humans from embryo to old age.

Of all animal species, humans have one of the largest increases in life span over our seeming genetic predisposition. We are also unique in our degree of behavioral plasticity and access to diet, lifestyle, and medical advice to lengthen healthful life. Researchers noticed that, as brain size increased, so too did the life span of early primates. Reproduction played a key role in determining mammalian life span. These traits seemed to be the major players in human aging just when Kenyon, Ruvkun, and others found the pivotal role of the endocrine system in worm life span. But which was most important for humans?

Slagboom spent twenty years answering that question. She cofounded the Leiden Longevity Study to research both the long-term health effects of the Dutch Hunger Winter and to collect samples of her long-lived countrymen, or "human longevity mutants," she called them. She collected the blood samples of 3,500 individuals, with an additional 500 families of nonagenerarian sibling pairs (brothers and sisters in their nineties), with one child and the child's spouse as a control. She found that her long-lived subjects had a healthy metabolism; less heart disease, arthritis, and diabetes; and greater resistance to stress. The most universal common element to healthy aging was "really the preservation of insulin sensitivity," she said. When Kenyon's and Ruvkun's experiments, and the following mice and fly experiments also pointed to insulin, "that directed our attention to the insulin/IGF-1 pathway," she said. Eventually, her team found a triangle or overlapping network involving insulin, fat metabolism, and the immune system. "Insulin sensitivity is the key," she said. "But it is a system, not one gene. TOR's signaling is also very important."

When she turned to the effects of the Hunger Winter of 1944, Slagboom found a deep effect on prenatal epigenetics, or the marks on genes. "We found scars, if you will, on genes which lasted into the people's sixties," she said. "If your mother is starving in the first three months of your gestation, you lose DNA methylation, a chemical change as if the life-span thermostat is shortened. In your middle age you become overweight and suffer hypertension." Slagboom found the key gene turned on in the womb out to be IGF-2, the second form of insulin-like growth factor. "A lot of other genes are involved too, some sixty in all. The system's not simple! Each element contributes, as do telomere lengths and mitochondrial health."

Around the world, the race was on to find the human longevity genes by researchers, some of whom jokingly called themselves FOXO hunters. FOXO proteins were a subgroup of transcription factors in the insulin signaling pathway, involved also in cancer. Slagboom's Dutch study found the *FOXO3a* gene in its nonagenarians. An Ashkenazi Jewish study out of the Bronx's Albert Einstein College of Medicine by Nir Barzilai showed that many centenarians shared the defective insulin-like growth factor receptor. Similar studies showed similar findings in Chinese, Scandinavian, and Italian populations of centenarians, along with other kinds of stress resistance and DNA repair. The gene appeared most prominently in the study of Hawaiians of Japanese descent. The key to *FOXO3a* seemed to be the specific environmental conditions that trigger it. "It's the most consistent validated gene in the field," said Albert Einstein's Nir Barzilai.

The single biggest behavioral trait contributing to human longevity, noted in the famous Stanford University seventy-year-long study, was "conscientiousness." By 2007, the MacArthur Foundation in Chicago thought the question was so urgent it funded a study of a potential 'longevity dividend" in human longevity, in opposition to the crisis many experts predicted. "The elderly have the lowest carbon footprint of any age group. They are not using economic resources and bring economic benefit to a community," said study group member Jay Olshansky. The fiery professor made headlines telling criticizing doctors: "American physicians give us the most resistance," he said. "Their business came from people being ill."

All of which brought back the interest of business.

Timing of Capital

In November 2007, the ideas of natural longevity, metabolic sensing, and human longevity genes were in the air as Sirtris published in *Nature* its discovery of two small-molecule activators of sirtuins, structurally unrelated to resveratrol and to each other, and supposedly up to "one thousand times" more effective than the natural compound. In December, the Sirtris video of the obese mouse running 40 percent longer after being fed resveratrol had become a hit on YouTube, and the *Wall Street Journal* ran a cover story on Sinclair and Guarente. The company's stock price jumped.

CEO Christoph Westphal had raised more than $100 million from investors. Back in May 2007, the initial public offering had been successful. Westphal had even held off investors asking when the next round of financing would be. In early 2008, the company was worth more than $100 million, at $8.85 a share.

In the spring of that year, as the company applied for FDA approval of human clinical trials of the safety of SRT501, its first artificial version of resveratrol that could be patented and sold, and Westphal fielded calls from several pharmaceutical executives interested in purchasing the company. The most insistent was Moncef Slaoui at GlaxoSmithKline, who thought his company needed some entrepreneurial boldness. Westphal played hardball. If GSK wanted their science platform, it would have to pay.

Some scientists at GSK, however, remained skeptical about sirtuins' role in human late-life health and the effect of resveratrol or the other agents on sirtuins. "Polyphenols typically show a lot of different effects, typically at the same time. They are very hard to develop as a drug. People tend to get phenolic hits and just go 'aaaah' and go on to someone else, because the success rate is so low," said Derek Lowe, founder of the pharmaceutical blog In the Pipeline. Competing scientists at Amgen and Pfizer began studies to see if they could replicate Sirtris's reported findings with resveratrol or the other "small molecule activators." Many people did not believe that sirtuin activators mimicked caloric restriction's effect on longevity. The *SIRT1* gene did not extend mammalian life span, though it was linked to improved metabolism in mice eating a high-fat diet. There were no other known activators of enzymes similar to sirtuins. Most drugs inhibit biochemical processes as antagonists, such as cholesterol inhibitors. It was technically difficult to make an activator. "People were suspicious," admitted Sirtris senior director of corporate development Michelle Dipp.

Sirtris was attractive to the pharmaceutical giant GlaxoSmithKline in part because the worldwide corporation was restructuring to be a collection of independent units, each with less than a hundred people. Sirtris could handle the discovery of compounds. It could do the early clinical testing. But it needed GSK's size and expertise to do the big human trials—the kind of huge clinical tests the FDA would require.

The turning point had been the second *Nature* article on the running obese mice treated with resveratrol. Media interest again perked up. The respected *Wall Street Journal* biotech reporter David Stipp, who had put Guarente and Sinclair on the newspaper's front page, explained the factors in Sirtris's favor:

> The resveratrol study had more going for it than any anti-aging effects I'd seen. First the work was blessed by *Nature*, and its authors worked at prestigious institutions. . . . [After visiting the federal lab in Baltimore] I came away convinced that the long, weird quest to extend life span—a five thousand year trek during which hopelessly hopeful seekers tried everything from transfusing blood from youths into their aged veins to injecting minced dog testicles—was finally getting somewhere.

Some at big pharmaceutical companies were looking at cooperating more with academics and start-ups that had the new ideas. Taken together, the Sirtris science papers outlined a compelling potential pathway to some in an industry looking for new approaches.

At the same time the company Elixir sought to go public itself. The pharmaceutical giant Novartis evinced an interest in Elixir, which had turned itself into a company fighting mainly late-life metabolic disease. Guarente had resigned from the company he cofounded and had moved to his former student's higher-glitz start-up. But Elixir had a good, high-profile platform to sell. It owned the rights to mitiglinide, a Japanese-developed glucose-lowering drug. It had a grehlin antagonist working on the insulin pathway and a sirtuin inhibitor for Huntington's disease. It had completed a third round of financing. But when the venture capital company MPM had taken over as the company's biggest shareholder its mechanism of business—late-stage targets with significant dollar value they could sell to pharma or take to the stock market—took precedence. There was pressure on Elixir to forget about longevity and focus only on drug discovery.

As the pressure mounted, Elixir prepared to go public. It wanted to use as a springboard a JP Morgan meeting in the beginning of January 2008. Then the stock market had a terrible January. No initial public offering got done for a couple weeks. Under the conditions, the company stock price was too high. By February 2008, Elixir had $50 or $60 million it could count on, but the company had to reassess itself. The FDA asked for several additional studies. The company did a fourth round of private financing, bringing in a new venture group, Hercules. It laid off several researchers. "Science is a moving train," the award-winning virologist Bernard Roizman once said. "You have to know when to get on and when to get off." By 2008, the stakes, money, and competition in the scientific and business race to extend healthful life had increased exponentially.

Four pathways emerged as front-runners, but the fortunes of those genes had as much to do with media coverage and institutional power as with scientific fact. In one sense, the longevity gene timeline was moving exactly as science innovation was supposed to, from pure research to professional discussion to private investment. It took a lot of money to move from the lab to the clinic, and an academic lab was no place to develop a potentially life-saving drug. Science, the most successful knowledge-generating system in history, was confronting the problems of society. The studies in the wild pointed to cell-stress resistance, insulin, and mTOR, coupled with longevity workhorses like telomeres and sirtuins. "It's a network," even David Reznick admitted. "What Kenyon and others are doing might yield a therapy."

The result appeared to be what research in a capitalist democracy was all about: risk, mistakes, passion, imagination, and hype. Then the whole thing exploded.

9 *The Rush and Crisis: 2008–2010*

In January 2007, bankers surrounded thirty-nine-year-old Christoph Westphal, as if he should have worn a "papal ring," said one Sirtris executive, when the company prepared to go public. In February of that year, the company had made the cover of *Fortune*, followed by profiles in *Discover* and *Newsweek*. By the beginning of 2008, Sirtris was courted by GlaxoSmithKline for what would be the biggest-ever purchase of a biotech start-up. These developments seemed to make a parable of pure science moving to a public health breakthrough in a profitable application of a lab discovery. The story had sold itself in raising some $113.5 million in five rounds of private financing and another $62 million in its initial public offering. "The antiaging message is very powerful," said Westphal, "especially when you are talking to a bunch of aging, overweight guys who are prime targets for the drugs you want to develop."

In 2008, the giant pharmaceutical manufacturer GlaxoSmithKline epitomized the term "multinational corporation." Formed in 2000 when British companies GlaxoWellcome and SmithKline Beecham merged, it had sales in 2007 of £22.7 billion. Employing a hundred thousand people in more than one hundred countries, it produced one-quarter of the world's vaccines and had discovered the antibiotic amoxicillin. But its director of research and development, Moncef Slaoui, a Moroccan with a PhD in molecular biology and immunology from the Universite Libre de Bruxelles in Belgium, saw innovation struggling inside a slow-moving organization.

At the time, new ideas at GSK required several levels of approval. The best way for company scientists to advance was to move into management or a new area of research. With a new CEO—Andrew Witty appointed in 2008—the company sought a more entrepreneurial model. Slaoui and Witty proposed that GSK foster small, in-house, independent units almost like start-ups, encouraging researchers to pursue personal, and profitable, interests. The leadership sponsored in-house academic seminars on cutting-edge science and then went looking for exciting small new companies to purchase.

Of those companies, few seemed more promising than Sirtris, with its running obese mice, glowing media reports, strong stock price, and the public embrace of resveratrol by Oprah's TV doctor Mehmet Oz. Academic researchers had shown that the mechanism by which resveratrol extended health was to increase the number and efficiency of mitochondria, the cell's energy factories. The company's first synthetic sirtuin activator, SRT501, was in safety trials in India. The market looked enormous. By 2050, 76 percent of the US government income tax receipts would be going to entitlements for some seventy-seven million–plus aging baby boomers. Their late-life health was critical. Slaoui called the sirtuin platform a "potentially transformative science."

Still, the start-up was at a very early stage. Polyphenols like resveratrol showed a lot of different effects. In the original 2003 experiment, Howitz and Sinclair had neglected to validate their results with a second, different test, often a radiometric assay. If they had, the experiment would not have worked. The sirtuin enzymes story, too, was complicated. They certainly had important biological roles, but it was going to be a long time before the company had an idea of what they were actually doing. Each one could be activating or downregulating hundreds of different genes or proteins, in different tissues, in different systems at different times.

Andrew Witty heard the concerns. "If we are wrong, we might get nothing," Witty admitted, "but this is exactly the type of thing a company like GSK should have in its portfolio," he said. Sirtris stock was trading at $12 a share, yet Witty agreed to pay $22.50 a share for the company. Sinclair's profit was estimated at $5 to $6 million. He continued as an adviser at $297,000 a year. Westphal agreed to direct a GSK investment fund and insisted the corporation give Sirtris employees similar lucrative options to those it gave him.

The biotech and business media applauded the GSK purchase of Sirtris. Favorable articles on the acquisition ran in the *New York Times*, *Forbes*, and in trade newspapers like *Bioworld Today*. Two Harvard Business School case studies appeared, one covering the Sirtris decision not to market resveratrol as a health food and the other writing about its loss of independence when it was purchased. Sirtris posted both on its website. "You could not buy better advertising," said Michelle Dipp, though one study cited a misleading statistic that the obese mice fed resveratrol lived 30 percent longer than normal. The figure was 12 percent. "Everyone wanted to believe in it. It moved so fast," she said. "We are the fastest company ever to go from a novel compound to man. We each feel that this will be the most important thing we do in our lives."

By January 2008, Sinclair was presenting results to Estée Lauder executives in Paris, whose company helped support his research. In April, GlaxoSmithKline bought the company. By June, Westphal had traveled to Switzerland to consult with European mega-insurer Swiss Re on the economics of longer life spans.

But the CEO remained cautious. People thought him a relentless promoter, yet internally he was constantly resisting the pressure to overpromise. "Part of my job is to downgrade the expectations," he said to a tech investment newsletter. "Being inside and just confronting the long time frames and the extraordinary failure rates was surprising. I didn't fully understand how difficult it is."

In a moment of science upheaval, the techniques of persuasion become pivotal, and a sense of caution and the absurd proved good qualities to have.

The Runaway Train

Sirtris's glowing media reports drove other researchers slightly crazy. "It is hard for the field to advance with misinformation hanging," said Cynthia Kenyon, referring to the growing body of evidence challenging some of the resveratrol and sirtuin claims. Even seasoned, respected journalists appeared to be caught up in the hype. "Why is Nicholas Wade in Sirtris's back pocket?" asked one researcher of the respected *New York Times* writer who wrote several prominent articles on the company. Peter Distefano, former chief scientific officer at Elixir, said, "You get punished for explaining your science correctly, with controls and counterexamples. Where was the peer review?" The coverage became, as Brian Kennedy put it, "a runaway train."

In June 2008, Harvard and Sirtris sponsored a Healthy Life Span conference in the New Research building's wood-paneled auditorium. I joined some 250 young people and researchers as they scribbled notes on laptops, graph paper, composition books, and reporter notebooks, checking scientific papers off the web as speakers cited them. "This is a very good day for sirtuins," said Lenny Guarente to me at the cocktail reception. Paul Glenn, the Harvard alumnus and investor who created the Glenn Aging Research Foundation in 1953, told the group: "GSK was the first major pharmaceutical company to make a major bet on aging. Now all major pharmaceutical companies are taking a look."

Still, problems faced the field. What was aging? How could it be measured? "We still cannot define it," said Tom Rando, the intense, dark Stanford University stem cell biologist who received the Howard Hughes and Paul Glenn research awards at the conference. "I call it a decline in structure and function. The cause is a network of influences and effects, with a tremendous amount of nonlinearity and feedback."

At the conference dinner, the scientists continued to argue the definition of aging. "You're an evolutionist, Marc, you tell us," someone asked Marc Tatar.

"That's why I'm not saying anything!"

"Maybe aging is some kind of systems failure," said Tom Rando.

"Mostly it is semantics," said Jan Vijg, a pioneering Albert Einstein College of Medicine chair of genetics and researcher in the epigenetics of aging, in clipped Dutch accent.

"You cannot study aging at just the cell level," said Rando. "That's why people got Leonard Hayflick's discovery wrong."

"Is there a mammalian equivalent of what's happening in worms?"

"There were a lot of contradictions in Rich Miller's talk! I did not agree at all," said Vijg about the Michigan researcher's claim that aging is not disease. "I can see lots of ways in which aging is a collection of diseases."

"What is disease?"

"Something your doctor can bill you for," Vijg said.

As summer passed, public interest and researcher zeal brought new difficulties. In August 2008, David Sinclair joined the paid board of Shaklee, a California company that sold Vivix Cellular Anti-Aging Tonic, which contained resveratrol. Structured like Avon cosmetics with independent dealers, Shaklee offered customers the chance to "feel 25 years younger." At a sales conference, Sinclair foresaw a time when "we can actually make a product that slows down aging," and on a Florida radio station with Shaklee's chief doctor, he repeated the claim that resveratrol increased obese mouse life by 30 percent. The figure was actually 12 percent, but Sinclair was calculating from adulthood to death. By December, under questioning from the *Wall Street*

Journal and pressure about a Harvard policy that professors should not endorse products, Sinclair resigned from the company, claiming it had misused his quotes. The company responded: "Every Shaklee use of his name—whether in print or in video—was pre-approved by him."

Still, public fascination skyrocketed. In January 2009, Sinclair and Westphal appeared on *60 Minutes* on CBS, and the hits on Sirtris's website rose from five hundred to fifteen thousand to twenty thousand a day. A spokesman for the biotech industry website Xconomy.com complained it was overwhelmed by callers seeking resveratrol. When Oprah's Dr. Oz praised resveratrol and acai berry as supplements, Internet marketers of the compounds used his name so much that Harpo lawyers filed a multimillion-dollar lawsuit against forty of them. "A confused 64-year-old" asked on the website: "So is Dr. Oz saying they're good or not?" Another complained: "I trusted Oprah and Dr. Oz."

As each newspaper, magazine, show, or website covered the story, others followed. In the same month as the Harpo lawsuit, Nicholas Wade wrote a well-researched article on Sirtris that received the front-page spread in the *New York Times* science section. It was titled, "Tests Begin on Drugs That May Slow Aging." The respected science author's profile cited the critics of the longevity claims. Yet the article was more important for the company than a *Cell* or *Nature* paper. (One study showed that when a scientist is quoted in the *New York Times*, her or his work is cited as much as 72 percent more often in professional journals than those who are not quoted.)

The public hunger, complicated biology, and deft showmanship turned science communication into a marketplace. "If you wanted a story to sell, this was it," said Wyeth's former vice president of metabolic disease and hemophilia research George Vlasuk, of the claim that a red wine compound extended healthful longevity. "It made a beautiful story, almost perfect. Christoph Westphal is a venture capitalist and did what a VC should. He played to the interest. The fact that the science lagged, didn't keep up with the claims, is the basis for a lot of the nonsense that followed." Hearing Westphal speak, Vlasuk became fascinated by the promise of sirtuins. Eventually, he would leave Wyeth to take over Sirtris.

Still, the sense of urgency grew. "People put their private fortunes into this," Sinclair said to me. "It's a big weight on my shoulders. I'm committed to making a drug. I'm worried about clinical trials. People's lives are at stake."

The Competitors

As the attention increased, the researchers into other longevity genes put their "foot to the pedal," as Andy Dillin said. In La Jolla, Dillin studied the role of reduced insulin-like growth factor in delaying or preventing Alzheimer's in mice. In Marin County, Gordon Lithgow found several promising hormone interactions based on the insulin pathway, some of which he felt could be ready for drug development. In the Bronx, Albert Einstein Hospital's Ana Maria Cuervo studied the role of insulin sensitivity in maintaining the cell's lysosome, its "garbage can that processes or recycles the cell's waste," she said, and moved up to codirecting the center's newly expanded international aging program. In Chicago, Rick Morimoto focused on protein maintenance and repair in the cell. Researchers searched about half a dozen molecular mechanisms, including protein folding, mitochondrial efficiency, and telomere lengthening, many circling back to the big four gene pathways—the sirtuins, *FOXO* and the insulin pathway, mTOR, and a new player, the fuel gauge AMP-kinase.

At Stanford, Anne Brunet studied the connections among these overlapping networks, focusing on aging and cancer, uncovering the links between FOXO3a, the human version of the Sweet Sixteen transcription factor, and the energy sensor AMP-kinase, with the health benefits of dietary restriction. In mammals, Brunet discovered that AMP-kinase controlled the life-extending transcription factor, and it seemed to play an important role in the sirtuin pathway as well. At the Jackson Laboratory in Maine, David Harrison and others found that rapamycin, the compound affecting the longevity gene mTOR, extended older mouse life spans by some 30 percent. By December, the Hawaiian researchers Bradley and Craig Willcox, along with Albert Einstein Hospital's Nir Barzilai, discovered that variations in the *FOXO* gene lengthened life in Hawaiians of Japanese descent and in Ashkenazi Jewish centenarians.

In 2009, the Nobel Committee recognized the biology of aging as a science by awarding its Prize for Medicine to Elizabeth Blackburn, Carol Greider, and Jack Szostak for their discovery that the enzyme telomerase protects chromosome endings. The committee called the discoveries important "pieces of the puzzle" of longevity. Several biotech and cosmetics companies sprang up to market telomere-lengthening products.

In San Francisco, Cynthia Kenyon's lab sought compounds that could affect the insulin receptor and *FOXO* gene in human cells. In Boston, Gary Ruvkun's lab found that the *FOXO* gene triggered in the worm body the genes that set its embryo clock to zero. Ruvkun had shared the prestigious Lasker Award for discovering the crucial regulatory role of micro-RNAs in the cell. He also won his NASA grant to create an instrument for detecting DNA on Mars. In Los Angeles, Valter Longo studied a group of small-sized Ecuadorians who lacked insulin and insulin-like growth factor, and who never got cancer or heart disease.

In higher animals like primates, the life-span experiments took a longer time. The University of Wisconsin–Madison's Richard Weindruch monitored his rhesus monkeys on caloric restriction for their twenty-fifth year. Featured on *60 Minutes* and in the *New York Times*, Weindruch's work suggested that caloric restriction extended healthful life, not just in mice and rats, but in primates too. A similar NIA study, however, was producing opposite results. Nevertheless, with associate Tomas Prolla, Weindruch had begun discussion to sell their company LifeGen to Nu Skin, an international skin-care company that wanted their longevity gene expression database.

The business interests, however, affected the science relationships. "We don't talk like we used to," Dillin said of his former Kenyon lab mates, several of whom lead new labs of their own. The language of business seeped into science conferences. At the Harvard Life Span conference in September 2009, for instance, medical school dean Jeffrey Flier's introduction echoed Christoph Westphal and the *Fortune* cover feature on Sirtris: "It is no longer a question of if, but when," Flier said of the prospects for longevity's medical applications. The conference was divided into two auditoriums: one of scientists, including Kenyon, Weindruch, long-time mouse researcher David Harrison, and others talking about their latest studies; the other featuring Sirtris-affiliated talks. The headlines the next day in the *Wall Street Journal*, *Boston Globe*, and *New York Times* mostly focused on the company. Save for an article by Nicholas Wade on rapamycin, the general science presentations were ignored.

Still, the scientific idea took hold. Genes influenced healthy aging by slowing it, molding it, postponing it, and by increasing the fitness of other systems. Four molecular pathways made overlapping networks modulating longevity in model organisms, each with many components. At long last, it seemed, a deserving new approach to late-life medicine was coming of age. But there, a land mine waited. Eventually, it detonated.

The Crisis

By the fall of 2009, the challenges to the sirtuin and resveratrol claims became too significant to be ignored any longer. Resveratrol or the other small-molecule activators did not trigger sirtuins, as reported. They triggered the researchers' fluorescent assay, or test. Tweaking sirtuins had not significantly extended healthful life in many mammal studies. The gene appeared to play a little, or selective or indirect, role in the mechanism conferring the benefits of caloric restriction. Consternation grew that such claims had appeared in journals like *Cell*, *Science*, and *Nature* and had gone uncorrected. "The problem with the original *Science* and *Nature* papers," said the University of Washington's Matt Kaeberlein, "was they were wrong. But because they were short, flashy, and lacking in mechanistic detail, it was hard to see. Smaller journals go through more rigorous peer review." Indeed, a 2011 study by two medical journal editors uncovered the fact that major science publications suffer from a fifteenfold-higher retraction rate than do lower-impact journals. This trend hinted at "the dark side of the hyper-competitive environment of contemporary science," wrote the study's authors.

By 2010, University College London fly researcher David Gems found that overexpressing the *dSir2* gene did not extend fly life span, nor did the *sir-2.1* gene extend worm longevity as reported. His colleague Linda Partridge recalled the reason for their experiments:

> While we were doing this [fly] work, the paper came out saying if you overexpress *sir-2.1* in the worm it extends life span. I thought that's funny, because we're getting negative results. So I got the original worm lines used to make the claim and found that as they were, indeed, they did extend life span. But if you did the proper controls, the results weren't there. I went up to David and said we've got to find out what's going on. Because . . . there's this huge rumor that we've got a major evolutionarily conserved pathway here, and I suspect that this is quite entirely wrong. I had to nag him for a very long time. But he

eventually did look at the worms carefully, and then he . . . did a very thorough job.

Gems and Partridge showed that the original effects had never been tested when you standardized the genetic background with the controls. If they repeated the experiment exactly as it was done in Guarente's lab, they got the reported result. But in England and the United States, some five labs found that if they did a standard genetic outcross, that is, mate the animals with nonmutants up to six times, the *sir-2.1* gene did not extend life span in worms or flies. The Guarente experiment had unwittingly allowed a second gene to alter the results. The labs tried to publish their paper but was met with initial rejection at *Nature*, where Partridge herself served on the scientific board.

Meanwhile, Matt Kaeberlein and Brian Kennedy confirmed that restricting calories extended yeast life span even in the absence of the *SIR* gene. The gene and the dietary effect acted on different genetic pathways.

In December 2009, Amgen pharmaceutical scientists announced in *Chemical and Biological Drug Design* that resveratrol did not activate sirtuin 1, the human version of the Sir enzyme. A month later Pfizer pharmaceutical researchers showed in the *Journal of Biological Chemistry* that neither resveratrol nor the "supermolecules, 1000 times more active than resveratrol," activated the SIRT1 enzyme. The original reported results were "a complete artifact," said Peter Distefano, meaning they were a ghost effect of the technology. Another key claim, that resveratrol-mimetic SRT1720 lowered glucose levels and improved mitochondrial function in mice fed a high-fat diet, was also challenged. None of the resveratrol mimetics was precisely selective to the sirtuins. Instead they hit multiple targets, meaning they "had no use as drugs," commented researcher Derek Lowe on In the Pipeline.

Sirtris executives portrayed the conflicts as arcane disputes among rival academics. At a Cambridge Biology Information Technology conference in April 2010, Westphal mentioned the "debate in the academic world" about the activation of SIRT1. He claimed that "everyone in the field agrees our molecules have beneficial effects in mammals." Sinclair wondered if the scientists were working properly with resveratrol or its mimetics and claimed the Pfizer scientists may have lacked expertise. GSK officials added that it had other activators anyway. Andrew Witty called the conflicting data a "tempest in a teapot. We're not at all surprised there's some controversy. Frankly," he said in an interview on the website FierceBiotech, "we didn't think what was published was particularly comprehensive."

Westphal's comment prompted Derek Lowe to finally air out the conflicting reports on his blog. An Arkansas native who earned a PhD from Duke in organic chemistry before working at several German and European pharmaceutical companies, Lowe had been writing his widely read but unpaid blog for seven years, because "between people who do this stuff and people can talk about this stuff, there's not much overlap," he said. In a commentary, "Everybody Agrees?," Lowe reminded readers that he had been a fan of Sinclair and the Sirtris approach since before 2003, but now questioned some of the claims. In January 2010, after reporting on the conflicts on two occasions, Lowe noted that Sinclair's response to the authors of conflicting reports had been "close to insulting." Lowe concluded in a later post of the Pfizer article:

> What we learn from this paper [by Pfizer] is that the assay is worthless for even more complicated reasons than originally thought, and the whole series of SRT compounds behaves in ways that were not apparent from the published work, to put it lightly . . . Pfizer's gauntlet is still thrown down right where they left it . . . What did GSK get for its $720 million?

Lowe's reports unleashed a stinging web commentary with all the urgency and unchecked emotion of any Internet reviewer thread. "GSK fired ⅔ of its cardiovascular research group, and one-half of their metabolic group!" wrote one commenter. "It's basic enzymology!" complained another of the lack of controls in the original papers.

The blog comments signaled the coming of a new form of science communication. Formerly the purview of a handful of national reporters, science now exploded onto personal websites written by insider, sometimes angry, lab benchers. At conferences one could see rows reserved for blog or Twitter commentary, as audience members analyzed findings the minute they were mentioned. Many web readers were huge science fans. "I'm actually quite excited about it [aging research]," Lowe concluded. "I just didn't want to lead people on. I didn't want to be led on myself."

Others commented in stronger terms. "Scientists are supposed to remain impartial," observed Cynthia Kenyon to me in 2008. "If there is a financial conflict of interest, then that needs to be disclosed. The public assumes that scientists generally do not report things in an unbiased fashion. It's a pretty serious thing in scientific circles." Probably some of the insulting tone of Sinclair's initial response backfired.

So too did some of the Sirtris executives' genuine attempts to improve people's lives. Because of the many unregulated resveratrol knockoffs being marketed, Dipp and Westphal had created the Healthy Lifespan Institute, a nonprofit organization that sold the pure compound "at cost," said Dipp. In August 2010, GlaxoSmithKline moved to prevent Westphal and Dipp from doing so, because it was "conceivably undermining Glaxo's nearly $1 billion investment," reported The Street. By then, the seasoned biomedical researcher George Vlasuk, whose specialty was vascular disease, had replaced Westphal as Sirtris CEO, as planned.

But the problems for resveratrol and Sirtris remained. By the end of 2010, GSK halted development of one Sirtris resveratrol mimetic called SRT501. The compound showed it could have a therapeutic application for myeloma, a bone marrow cancer affecting more than sixty thousand Americans a year. However, test subjects became dehydrated and nauseous, as had the rhesus monkeys taking the compound before them, and at year's end all further work on the compound was ended.

By the time the dust settled, it was questionable whether there was much significant, repeatable activation of the human SIRT1 enzyme by resveratrol or by its second generation superversions. "The science was not as developed as it should have been," said Vlasuk, who quieted the company's pronouncements and set out to disprove the Pfizer claims, while directing new research in new compounds. "The SIR bubble," David Gems noted, "burst."

An Idea's Power

The controversy did little to slow the public hunger for resveratrol. Indeed, many cancer patients commenting on the myeloma website asked for studies to continue on the natural version of the compound. The demand grew more insistent.

The effect of the red wine story was like that of Midas's gold or Tolkien's ring. People wanted healthful longevity at any cost. Reporters had a duty to inform the public. Science was a form of theater, and teaching it was a form of selling. As the attention, money, and stakes increased, the passions and plot twists deepened. Aging was complicated, and cause and effect at the molecular level are difficult to pin down. Media coverage made a chorus that magnified early mistakes into a grand opera beyond that which was originally intended. Sinclair complained that "if I tell an interviewer there's 'a chance of life extension,' the next day the headline reads, 'Sinclair promises life extension.'" Peer review at some major journals faltered. Profit made a driving force in the halls of pharmaceutical science. Some web commenters lambasted GSK's unwillingness to listen to its in-house scientists. A few called for retractions.

Yet a huge investment in a questionable premise was nothing new in research. "It's not a crisis," commented Linda Partridge. "It's what science does." She spoke to me from Cologne, Germany, where she had moved to direct the brand spanking new Max Planck Institute for the Study of Aging. Church bells tolled in the background. "The *SIR* gene and resveratrol do a lot of things, some very good." she said. A follow-up experiment by Helfand seemed to show that overexpressing the gene did extend fly life span.

For all the controversy, aging study had publicly altered in its scientific definition. Aging was no longer a random accumulation of injuries and mutations as previously thought. "There's a lot of potential there," concluded Lowe. "The evolutionary distance between a fly and a human is less than that between a fly and a roundworm. These longevity mechanisms are likely to be highly conserved. There's clearly room for human longevity to be altered. But I don't want to be the first to take a drug!"

As other nations saw what was happening and joined in, the race accelerated. Aging research in Europe expanded. China invested heavily. Brian Kennedy (now president and CEO of the sleek Buck Institute for the Study of Aging), with Matt Kaeberlein (now codirector of the University of Washington Medical School Nathan Shock Center for Excellence in the Basic Biology of the Study of Aging) and Albert Einstein Hospital biologist Yousin Suh, accepted positions as distinguished visiting professors in China at Guangdong Medical College's Aging Research Institute. "They're a little behind," said Kaeberlein, "but moving incredibly fast." The driving forces were China's aging crisis, one of the worst for any country, and conventional market capitalism. "On the street," Kaeberlein recalled, "the peddlers were trying to sell me DNA!"

Four longevity candidates continued to run neck and neck as their study was liberated, it could be said, from some of the distortions of institutional power and financial or news hype. Resveratrol had brought unheard-of attention to the biology of aging, which was expanding with new excitement, tools, and imaginative approaches. The dream remained to turn pure lab discovery into a medical application to improve the future. Risk remained a factor. Even as doubts were raised, the interest in the new science increased.

The biology of aging made a medical problem, but it was also a social, economic, cultural, and even a spiritual problem. The research story offered a parable of science, but to understand its meaning fully, one question remained: Could any one of the lab discoveries be applied to treat just one disease of aging?

10 *Live Long and Prosper: 2009–2010*

December fog hugged the Pacific coast as I drove to one of the new labs in the study of longevity genes. Andy Dillin, now Salk Institute investigator and the new director of a Paul Glenn Center for Aging, had uncovered a compound that protects against Alzheimer's, which led him to join two other researchers and found his own biotech company. "I'm into promoting youth," he said. On the wall of his medieval-looking office hung a humbling Nietzsche quote his mother sent him: "You moved from worm to man but much within you is still worm."

By 2010, a new generation of researchers, including Dillin, Coleen Murphy, Heidi Tissenbaum, Anne Brunet, Malene Hansen, and others, competed to understand the ways in which a gene discovery might lead to a treatment of a disease. Is it a coincidence that the rates of Alzheimer's, cancer, and heart disease all rise with age? No: aging makes us susceptible to illness. The question was whether a longevity gene discovery would yield a treatment. To answer that, they had to change their focus. Rather than asking why we age, which had an almost infinite number of answers, it would be better to ask why we remain healthy as long as we do, which might lead to a few real, working therapies, or at least a promising avenue to explore.

Such applications of basic science required a grueling, boring industrial-type sift through tens of thousands of compounds. A lot of companies were trying, including perhaps Merck, Pfizer, Amgen, GSK, and Novartis, but they were not talking. The people who talked remained the academic researchers, moving from the familiar pure lab to the unfamiliar messy clinic to apply their discoveries. "We've become Rapa-land," the University of Texas's David Sharp science writer Gary Taubes for *Discover* magazine, referring to his institution's complete focus on the compound rapamycin. Such was the new phase of the biology of aging, pushed by investors and government agencies and, most of all, by the public.

Several avenues offered possible applications. Protein shape, cell defenses and repair mechanisms, patterns of healthy gene expression over a lifetime, telomere lengthening, isolating and removing damaged tissues, utilizing the insulin receptor or *SIR* gene pathway—each was promising in its own way. The challenge was to translate the basic science to patients in the clinic. To do that, one needed to make a high-risk bet on the right mechanism.

Another possibility was almost the opposite: not to improve cell defenses but to reduce them. "Ask any gerontologist to name the top aging diseases, and inflammation will come up in every case," said Judith Campisi. The defensive mechanisms that protect us early in life may later lead to the rise of cancers and heart disease. Campisi's and other labs screened for compounds to prevent dying cells from sending out inflammatory signals. Others sought to eradicate those damaged cells altogether. Some studied cancer-free wild animals for clues. Others looked into the mitochondria, the cells' energy factories.

Most every previous gene discovery remained on the table. If sirtuins had proved anything, it was that companies and private individuals were willing to pour hundreds of millions of dollars into the biology of aging, if one could sell a method of fighting the chronic illnesses of a developed world. "All longevity genes," said Gordon Lithgow, "are disease-fighting genes. We should not be thinking of limitations, but of new horizons."

Yet for all the promise and investment, twenty years of research had produced very little of therapeutic value. One rare cancer succumbed to one fascinating treatment, which we will study. But overall it was hard at Dillin's lab in December 2010 to avoid the conclusion that longevity biotech had so far proved a bust.

Above me loomed the twin white monoliths of the Salk Center, while below, the eternal waves pounded the La Jolla cliffs. I repeated the mantra: aging is not programmed. Or, maybe it is in simpler organisms. Genes influence aging, not by producing the aging state but rather by slowing it, molding it, trying to postpone it or improve the fitness of other systems. The processes were not chaotic and random, just complicated and intricate. It all depended on how we understood them.

As I waited for the elevator, I wondered which system might lead to a drug. Was it proper protein folding, cell defense and repair, something in a vast new field called epigenetics that looked at how genes changed over a lifetime? The steel doors opened. I stepped in, ready to look for answers.

Protein Folding and Heat Stress

Restless in his corner office and sitting beneath a three-foot-high, framed *Nature* cover featuring one of his articles, Andy Dillin eyed his touring bike in the corner. Dillin was looking older, a bit careworn in blue jeans, blue work shirt, and black shoes. He was probably thinking something similar about me. When we first met in 2002, he was a postdoctoral fellow and bike enthusiast in Kenyon's cramped lab. Now he directed the Salk's Paul Glenn Aging Center. Like Sinclair, Kennedy, Kaeberlein and others, he had advanced to the top of science in a field that once seemed such a backwater. "All of my work is on protein folding," he was saying. "If we can target the proteins folded by genes, we can target the diseases of aging."

Proper protein shape is a key to late-life health. In each of our trillions of cells, tens of thousands of proteins make beautifully molded pieces, each sculpted to fulfill a function. Proteins fold in a fraction of a second. In biology, form makes function. Stresses like heat, mutation, oxidation, and aging damage or denature proteins, destroying their form. The clumped old proteins clog signaling pathways. Damaged proteins play a role in aging diseases from cancer to Alzheimer's and Parkinson's, and perhaps most significantly in heart disease.

The main preservers of protein shape are the heat shock proteins. These protective mechanisms increase in number when a cell is exposed to heat or other stresses like infection, inflammation, toxins, or aging. Such protectors play a critical role in muscle development and recovery after a heart attack or a stroke. Moreover, the speed of recovery from heart attack could be enhanced by overexpressing heat shock proteins.

Invitingly, protein protection made one of the key mechanisms modulated by the insulin and *FOXO* genes that Dillin helped investigate in Cynthia Kenyon's lab. Protein folding thus made one of the useful potential drug targets, so important that the role of heat shock proteins in cardioprotection and neurodegenration was considered "one of the most important future directions in biomedical research," said

Northwestern University's Rick Morimoto, who was seeking "genetic and small molecule strategies to enhance protein quality control." Maintaining the proper folding and unfolding of proteins made a vital cellular activity in every organism from bacteria to humans and a key to the prevention of Alzheimer's. The research was going ballistic.

Across the hall, Dillin's postdocs worked rapidly in rows of narrow benches. Each bench featured a Nomarski-brand microscope, stacked petri dishes of worms, lab journals, and equipment-stuffed shelves, with the mice in a separate facility. As I toured the lab, some paused to speak with me. Kristan Steffen walked outside, shivering in the La Jolla December in her flip-flops, jeans, and cotton peasant shirt. Steffen had grown up in South Dakota and went to Brian Kennedy's lab at the University of Washington for her PhD. She now studied the ways in which the TOR gene kept proteins healthy. "I'm looking for the tissues where TOR translation takes place," she said. Two competing ideas explained where. "First, TOR could turn down translation in cells; therefore fewer proteins can misfold. The Buck Institute's Pankaj Kapahi's fly work suggests this," she said. "The second is that TOR is working on a subset of protective RNA, making sure it is translated better to protect proteins."

An *Esquire* feature described Dillin as the "most successful young scientist in the world." He had two major papers about to appear, one in *Cell* on metabolism and aging and the other in *Nature* on genes that modulate dietary restriction's effect. He had an Alzheimer's compound and a new company. But he felt unsettled. He had not taken a vacation in five years. A bulletin board poster charted the decline in research funds for aging. *SIR* was under assault. "If *SIR* had worked, it would have had a magic effect," Dillin said. "Because it didn't, we're bearing the brunt." He had dozens of competitors, each as ambitious and hardworking as he.

To relieve the stress, his lab members escaped on group outings, to the Del Mar racetrack, to a rock concert they nicknamed Worm-A-Palooza, and to house parties, like the James Bond dress-up party planned for the coming Saturday night. Their Stanford competitor, Anne Brunet, regularly held dress-up parties, like one for Tupac Shakur, or for Scorpio, her astrological sign, or for Bastille day and celebrations of her *Nature* papers. When I asked Brunet how many hours a week she worked, she demurred. "It's embarrassing," she said. Sixty? I asked. Seventy? "More," she said. The potential of pure research tantalized, but capitalizing was proving harder than anyone had imagined.

Others in the Dillin lab worked on other aging diseases. The tall, London-raised Will Mair studied the energy regulator AMP-kinase, had a forthcoming *Nature* paper, and was applying for jobs. The soft-spoken Californian Erik Kapernick, in mismatched shirt and pants, worked on Alzheimer's disease studying the beta amyloid protein misfolding that befuddled memory. A classical musician, he relaxed by teaching piano after lab hours. "There are so many connections between music and science," he said. "Genetics is like a symphony. We thought it was just one gene somewhat naively. Aging is a network of feedback loops in humans. Pattern, rhythm, and rules drive them. The genetics of aging takes time, which is one thing the universe has plenty of."

The trouble was venture investors and policy makers did not feel that way. The clock was ticking on all the longevity labs and the press release promises. To understand the difficulty, it would be helpful to review how a drug is made, from discovery in a tiny animal to multimillion-dollar marketing campaign.

To Make a Drug

The process of applying a pure lab discovery to the medical clinic is called translational science. Because of the huge private investment in the new genomics, translational science had become the rage of our post-genome era. Connecting the basic bench discoveries in laboratories like Dillin's to patients in the world, it had to translate those discoveries into treatments. Translational science thus links expensive laboratories with the society that funds them. Ideally, it makes drugs.

To close that gap required special skills. You had to appreciate both ends of the biomedical spectrum. Translational scientists used tools from clinical medicine while keeping abreast of the latest news in basic sciences like molecular and cell biology, physiology, and pathology. They moved from tiny molecules to cells to organs and, finally, to whole patients. Their research could take decades and cost millions.

With the tantalizing longevity gene discoveries, people focused on the four obvious pathways—insulin, sirtuins, mTOR and AMP-kinase. But by 2010, an astonishing 247 known or suspected longevity genes in humans had been claimed. Perhaps half of our 22,000 or so genes might play some role in preserving a healthy life span. That idea did not include the many other aging or life-span theories vying for acceptance. To make some sense out of the chaos, first you had to choose one promising lab discovery. Then you looked for molecules that tweak the promising pathway. Then, you tested them in living organisms.

Once you understood some of the biological processes involved in a disease, you would find people who knew how to make drugs which could be patented and sold. Many of these are structural biologists, who design drugs to bind to a molecule that affects a biological process. A drug taken by mouth is absorbed through the intestines, from there entering the bloodstream. A drug molecule is small; if a protein is the size of a human being, then the drug molecule is about the size of their smallest fingertip, so it can attach to a specific spot on the large proteins maintaining health in a cell. Such a molecule attaches to a protein by fitting to a surface gap. Changing a protein's form changes its function. Structural biologists use methods like X-ray crystallography and nuclear magnetic resonance imaging, (an MRI, as you might get for a bad knee, but at the level of the cell), to build three-dimensional models of proteins, gaining insight into what type of drug molecule might bind to them. You hope the molecule is safe.

There were lots of new ideas about how to make drugs, including gene therapies and personalized drugs created from our individual antibodies geared to each of our own maladies, but few had yet panned out. Ninety-eight percent of human proteins had never been targeted. But if just one longevity gene discovery could yield one disease treatment, society could save enormous amounts on health-care costs, and many of us would worry less about diabetes, cancer, heart disease, and dementia. Several companies sought just that.

Of those companies, GSK/Sirtris dominated the field. Skipping past the arguments about the bench science, it focused on sirtuins' roles as metabolic regulators. The company had three drugs in clinical testing. The drug furthest along, SRT2104, had in preclinical tests shown it reduced insulin resistance and prompted glucose uptake. Having passed phase 1 safety trials in people, SRT2104 was being tested in phase 2 efficacy trials for diabetes, psoriasis, and vascular disorders. Sirtris's second compound, SRT2379, showed efficacy in multiple animal models against diabetes and inflammatory diseases. It had completed phase 1 safety trials in humans. A third drug, SRT3025, for neurodegenerative or metabolic disease, had also completed phase 1 safty trials. In obese mice it was increasing fatty acid oxidation, as had SRT2104. "These are early compounds, used to understand SIRT1 activation," cautioned George Vlasuk. "They may not turn into drugs. They are the first case studies. Our hope is that they will give us an early understanding of the biology." In 2010, company scientists published a paper in the *Journal of Biological Chemistry* challenging the claim that its resveratrol mimetic had not activated the sirtuin pathway.

Insulin remained a potential game-changer for late-life health. Glucose was the fuel, and insulin distributed the fuel. One modifier of insulin signaling, the diabetes drug metformin, was being tested in more than 1,100 clinical trials for obesity, cancer prevention or treatment, and heart disease. Johnson & Johnson studied it as a treatment of children suffering from a weight gain induced by ADHD drugs. At the University of California at Irvine and the University of Arkansas, trials of metformin were beginning for the prevention or treatment of colorectal cancers in those at risk. Mt. Sinai Hospital in Montreal, Canada, and others were testing metformin as a breast cancer preventative and treatment. Baylor College of Medicine studied it for prostate cancer. Several European institutes focused on the drug's activation of AMP-kinase and inhibition of TOR.

Several other companies chased other disease applications of the discoveries. Novartis opened an aging division to see if it could adapt the insulin receptor research to a drug for a disease of aging. Siena Biotech purchased Elixir's sirtuin inhibitor for the treatment of Huntington's disease and had it in phase 2 clinical trials in Europe. The company Proteostasis was developing Dillin's compound that instigated an Alzheimer's protective pathway, with similar discoveries by Rick Morimoto and the Scripps Institute's Jeffery Kelly. Nu Skin completed its purchase of Rick Weindruch's LifeGen for the screening technology to understand gene expression at different stages of life.

None of these studies were about aging, per se, a comedown which was inevitable. No country's drug administration would consider aging a disease. Indeed, the most surprising thing about gerotech in the twenty-first century could well have been the healthy amounts of money raised, jobs created, and research forwarded, just to treat something most people through most of history considered perfectly normal.

Despite the efforts of pharmaceutical development, historically, some of the most successful drugs remained those found by serendipity. Penicillin came from bread mold. Aspirin came from willow tree bark. The vast male erectile dysfunction industry originated with a failed Pfizer blood pressure medication in northern Scotland. True, some important new drugs had been designed using pathways discovered in the lab. Boston-based Vertex Pharmaceuticals, for instance, used rational design to create the hepatitis C drug telaprevir in May 2011, initially among the most successful drug launches in history. Merck, Millennium, Johnson & Johnson, and others had their share.

The biotech dream continued despite the faltering economy. A Chinese company invested heavily in the search for centenarian genes. In the United States, biotech start-ups raised some $22.1 billion in the first six months of 2011, The stock indexes still performed relatively well in a terrible market. Private-sector investment grew, raising some $2.9 billion in the first half of 2011. But a good deal of investment was shifting overseas due to American FDA regulatory hurdles, and there was a low level of early-stage investment. By 2012, the biotech market worsened considerably for new companies.

Still, a drug to treat a disease of aging, based on one of the known longevity gene pathways, could be hugely profitable. Yet more researchers, with more approaches, raced each other to find one. A place where people might have looked, given the clues, was in the wild. In Africa, a familiar small animal offered a new theme to finding a drug to improve human aging.

Cell-Stress Defense

In the 1970s, a South African undergraduate named Rochelle Buffenstein was hired to an underpaid, difficult job studying scrawny animals, "tubes with jaws," she called them, in Kenya's Tsavo East National Park. There were problems. The animal colony lay beneath the park's single main highway. "We had to break each time a truck passed," she recalled. Their other site was worse, on the road to Somalia, where an armed convoy dropped her in 1980. Little did Buffenstein know that this chancy first job would make her an academic leader in the race to extend healthful human life.

Seven feet below the parched savanna, the long-lived African mole rat introduced Buffenstein to aging far from the gleaming European and North American laboratories. The animals lived extremely long, healthy lives relative to their body weight, thirty years or more, the equivalent of humans living to a hundred years or longer. Focusing on the animals' remarkable societies, Buffenstein found that social cooperation helped their survival in the harsh environment. They had high levels of oxytocin, the breeding hormone, as a large number of males and females cared for pups. They never got cancer or heart disease. They did not feel pain. Their bone density, body mass, body fat, and metabolic rates remained youthful, and they reproduced until they died. They resisted the chemicals that caused free radical damage.

Living in colonies of three hundred or so, the mole rats built maze-like tunnel cities with separate rooms for food storage, resting, and ventilation. Most colony members were nonreproducing subordinates, with one queen reproducing with two or three males. The nonbreeding males had low levels of testosterone but at any point one could become a consort to the queen, and any female could become a queen.

Taking a colony with her to the University of Texas, Buffenstein realized that the animals' longevity paired with their other, more vital adaptations, including intelligence and social cooperation. "Longevity itself was not adaptive," she said. "It was linked to other traits that were." The animals felt no pain as a result of the low oxygen levels in their tunnels, and their high oxidative stress resistance she thought came from living in her San Antonio laboratory. They resisted the oxidative damage just as they resisted toxins or stresses of the wild. More important, even when injected with precancerous tissue, they simply avoided the disease, which was otherwise the prime killer of their rodent cousins. But how did they do it?

Buffenstein studied intently this cancer resistance. When an early tumor was detected, she learned, the animals' surrounding cells went into overdrive to destroy or neutralize it. "High rates of cytoprotective, detoxifying mechanisms or agents called cytokines took charge," she said. Cytokines are triggered by the Nrf2 gene, which produced a transcription factor that turned on the familiar protective heat shock proteins. In humans, cancer usurps Nrf2. But in the mole rats, Nrf2 arrests the cancerous cells in a static, nonproliferating state. "Nrf2 makes a gatekeeper," Buffenstein concluded. From her vantage, mammals could evolve protective mechanism from stress.

Other researchers looked at the same disease from the opposite angle, at the cancer-causing misfiring of human cancer defenses as we age. Judith Campisi had wondered, why does cancer increase with age? In the famous 1996 *Cell* issue that featured Kenyon's "Ponce d'elegans,"

Campisi observed a contradiction to the Hayflick limit to human cell life in culture: human cancer cells in a petri dish live forever. "I was convinced that the processes of tumor suppression would tell us something fundamental about life span," she recalled. "There was a basic cellular mechanism linking cancers and the cell senescence which prevents cancer and aging." But few noticed.

Cancer is rare in young people because of tumor suppressor mechanisms that corral and attack damaged cells, killing them with swelling and inflammation with those same mole rat cytokines (messenger proteins calling for more responders) and enzymes. As humans age, though, the repeated actions of these mechanisms, like waves against the La Jolla cliffs, build damage to the point where that protective response misfires.

These inflammatory systems likely helped us fight pathogens in the wild. But in modern culture they hardened arteries and developed diabetes' insulin resistance. High levels of inflammatory C-reactive protein (CRP) predicted heart attack more accurately than even high cholesterol or blood pressure. In response to plaque build-up on artery walls, the fat-eating macrophage attacked fat deposits, swelling and destabilizing them. Those deposits break loose into the bloodstream, causing clots and heart attacks. But no one had shown how to protect against such damage in live animals.

At the Mayo Clinic in Rochester, Minnesota, in the late 2000s, two researchers answered that challenge. In older patients, they noticed, dying or dead cells damage living tissues. The cells were supposed to die as part of our natural protective mechanisms. But as we age, we do not clear them out as well as when we are young. Could these senescent cells, secreting those same enzymes called cytokines that protected us when young, drive the damages of aging?

Working at first with little funding, the Dutch researcher Jan van Deursen and young American Darren Baker discovered that a mouse's dying cells set off a biomarker called $p16^{Ink4a}$ in animals they designed to age prematurely. They sought to hijack the communication system by designing an artificial gene they called INK-ATTAC, induced when they gave a specific drug. The mice with INK-ATTAC remained healthy. They did not develop the usual cataracts and muscle and fat damage as their untreated peers. They could run much further on a treadmill. Then the researchers gave the drug to adult mice. These animals' fat and muscle tissue damage was still delayed. The mice did not live any longer than the controls, but the researchers had shown the importance of dead or senescent cells in aging. "It really is quite a breakthrough," Judith Campisi told the *New York Times*. "We now know that there are processes driving aging, and that those processes can be meddled with."

When their *Nature* report appeared in November 2011, the researchers became brief celebrities, as had many longevity scientists before them. "Cell Study Finds a Way to Slow Ravages of Time," said the headline in the *Wall Street Journal*. "Prospect of Delaying Aging Ills is raised in Cell Study of Mice," said the front-page *New York Times* article. "It was highly unusual to put a mouse finding on the front page," said Nicholas Wade, the article's author. "I would think it was because of the completely new approach to the biology of aging." A Chicago-based venture capital firm approached them about starting a company.

These and other discoveries sought to turn aging biology away from hardwired genes of animals to the processes repairing and activating genes as we grow. Every day, our genes experience a daily flux of assault and damage. Most of these lesions are repaired instantly, but the protective and repair mechanisms could be damaged by diet, environment, stress, and age. There was a fancy term for this complicated area of study: epigenetics, Greek for "around the gene." Epigenetics in the new millennium's second decade became longevity's new edge. Perhaps diet, good living, and exercise could level the playing field of our genetic fates. "The potential is staggering," *Time* magazine intoned. "The age of epigenetics has arrived."

Epigenetics had been a long-studied field in cancer and, in the new millennium's second decade, it vied to become longevity's new edge.

The Epigenetics of Aging

The chronic diseases of Alzheimer's, heart disease, and cancer were once considered mainly diseases of gene mutation. Yet some percentage are diseases of epimutation, the changes in gene wrapping from thousands of daily cell damages affected by our environment and daily activity. If you get sunburned, smoke, eat a charred steak, get depressed, drink too much, or experience combat stress, you can damage your DNA wrapping, either the histone tails of the discs around which DNA coils or the shrink-wrapped chromatin that includes the DNA and histone proteins. These tightly bound supercoils are what fit DNA into its tiny space in the cell. Despite the fact that DNA is incredibly durable, its protective clothing is actually in a constant state of assault and repair.

Such epimutations happen in one of three ways. The agent can be a writer (add modifications), eraser (remove modifications), or reader (bind to the chromatin). A common epigenetic change is called methylation, gaining a methyl molecule on the ending of a DNA strand. Adding a methyl molecule turns a gene off, generally, and subtracting one turns a gene on. DNA methylation is important in normal growth. It is what tells any organ cell, say, a pancreas or liver cell, to become what it will be. The loss of DNA methylation is part of cancer's armament first noticed in 1969, and also a part of atherosclerosis, Alzheimer's, and immune system disease like Crohn's disease. The loss of DNA methylation, wrote the University of Copenhagen's Trgyve Tollesbol, made a "critical risk factor contributing to chronic age-related pathological states."

The other two main epigenetics alterations occur physically in the gene wrapping, either in the histones or chromatin structures. As we age, damaged histones generally lose an acetyl group, which is what made the discovery that Sir was a histone deacetylase so exciting. Chromatin structures could be altered in several different ways.

No one disputed that such external damage could be significant, though it was hard to isolate and study with controls in cells or animals. The critical question was whether the pattern of such epimutations could be passed on to future generations. Would it matter to you if your grandmother smoked or your grandfather suffered from posttraumatic stress disorder? And what did that say about a potential health-extending drug?

Certain important clues came from the Dutch famine sufferers and their offspring being studied at the University of Leiden. The famine victims suffered from methylated genes in their blood, primarily in the gene that coded for the hormone insulin-like growth factor 2. They seemed to be programmed to draw the most growth from the fewest calories and thus suffered higher rates of chronic diseases like diabetes early in life. Amazingly, these people passed on this "thrifty" epigenetic pattern, set in infancy or the womb, to their children. A similar clue was seen in prenatal studies of obese mice. For most of us, thankfully, the seed cells of sperm and egg are wiped clean of the epigenetic marks our parents and grandparents incurred in their lives.

Epigenetics made a wide-open new field, with tens of thousands of gene scars noted in aging lab animals but little initial understanding of their significance. New companies like Epizyme and Constellation sought to mine such changes for drugs to treat cancer, diabetes, and other chronic diseases.

The theory was that loss of DNA methylation as we age leads to "an increase in oncogene expression and other disease processes," reported Holly Brown-Borg, who studied her healthy, long-lived dwarf mice for clues to their low rates of cancer. Her lab found very high levels of DNA methyl transferase genes producing the enzymes that protect gene silencing, and arranged to collaborate with a Scottish bioinformatics laboratory to understand more. By 2012, the epigenetics company Constellation had entered a strategic partnership with Genentech to develop its discoveries.

Some researchers, however, like evolutionist David Reznick, questioned the premise. "It seems very fatalistic," he said. Others saw thousands of epigenetic marks in aging animals. Initially, Cynthia Kenyon called the field "a bit mushy." Plant and animal breeding, pointed out Gary Ruvkun, had been going on successfully for thousands of years as proof of *hard* gene regulatory pathways manipulated by thoughtful study. The original idea of inheriting acquired characteristics had a checkered past going back to the botanist Jean-Baptiste Lamarck and the folk wisdom of history. It had an intriguing element of truth but fell into disrepute.

In 2012, the most-studied disease of aging epigenetics was cancer. Interestingly, if the Dutch famine struck a pregnant woman in the second trimester, her child was more susceptible to breast cancer, if in the first, it was heart disease and diabetes. Carcinogens like smoking and radiation altered epigenetic regulation. In cancer, our stem cells aged much faster than normal, because they work overtime to repair the rapidly proliferating damaged cancer cells. The stem cells of a sixty-year-old cancer patient look as if they are two hundred years old, said the Anderson Center's Dr. Jean-Pierre Issa, a pioneer in leukemia research, who saw a clue in epigenetics to treating one rare late-life cancer called myelodysplastics syndrome (MDS).

MDS is a cancer almost exclusively of people sixty or older. Lifestyle and environment could alter the DNA methylation of genes to make them precancerous. Issa discovered the drug decitabine, which changed DNA methylation and attacked MDS by restoring the instructions of the cancer cells. "Cancer cells start as normal cells," he said. "We were trying to get the cancer cells to go back to being normal by reactivating genes that had been silenced." By 2011, after three phases of clinical trials, the FDA approved decitabine as the standard of care in the disease, a first win in the longevity gene concept.

Much of the new aging epigenetics work involved the sirtuin genes. David Sinclair, among others, proposed that "response the cell has to DNA damage is probably more important to aging than the actual damage itself." He and others researched the modifiers that repair DNA breaks. "When you're young, you have a nice youthful pattern of DNA repair," Sinclair suggested. "But as you get older, this pattern breaks down." At the University of California at Irvine, fly researcher J. Lawrence Marsh studied sirtuins' epigenetic effects on Alzheimer's. At Stanford, Katrin Chua studied sirtuins' epigenetic effects on cancer, as did Lucia Altucci at Italy's Seconda Universita degle Studi di Napoli. At Harvard, Raul Mostoslavsky researched sirtuins' role in DNA repair and metabolism, by focusing on SIRT6, while Marcia Haigis focused on caloric restriction and mitochondrial function. At the University of Frankfurt, Stefanie Dimmeler studied the role of SIRT1 in cardiovascular disease. Companies like Novartis, CellCentric, and Acetylon expanded their epigenetics programs. The potential was to understand "a fair amount about the compounds that control the chromatin marks, and we already have reasonable compounds that can inhibit or activate them," said the University of Pennsylvania Medical School's Brad Johnson.

Another key contribution of epigenetics could be in understanding stem cells' roles in preserving or restoring late-life health. Epigenetic marks guide embryonic stem cells into the tissue-specific stem cells they become. Anne Brunet explored the DNA stresses undergone by neuronal stem cells as we age, which brought her back to the insulin pathway. Her lab began testing compounds to treat the decline of neuronal stem cells. At the same time, Stanford's Tom Rando studied stem cells' ability to rejuvenate muscle and brain tissue by reprogramming adult stem cells to be more youthful. Several other researchers at the University of Wisconsin and elsewhere sought to understand if we may do anything to manipulate those repair and defense mechanisms in our lives. Then came a completely new approach.

A New Candidate

Despite the many setbacks, the march to longer life span continued as the biology of aging's potential treatments or applications multiplied. At MIT, Lenny Guarente found that overexpressing the *SIRT1* gene in mice made them resistant to Alzheimer's. A growing number of disease therapies seemed to be promised in the insulin pathway, including diabetes, heart disease, obesity, dementia and Parkinson's. The Danish government began the largest epidemiological study of natural resveratrol's healthful effects on a human population. In August 2011, the NIH's Rafael de Cabo found that obese mice fed resveratrol mimetic SRT1720 lived 44 percent longer than obese mice not fed the compound, a discovery published in a new small journal called *Scientific Reports*. Shortly thereafter, De Cabo made a similar finding in a small group of aging obese people.

After roughly twenty years, the longevity gene story made an explosive example of the new junction of capitalism, media and ideas in life science. As the gene discoveries piled up, the entire approach to ameliorating disease came under reconsideration. Traditionally government funding agencies like the National Institutes of Health attacked chronic illnesses individually, as if these diseases occurred independently of each other or regardless of the age of the individual. But interconnected degenerative diseases are the dominant causes of late-life illness. Despite the enormous expenditure of money in the last decade, many of the other high-stakes searches for new disease drugs had foundered.

As I drove back on Pacific Coast Highway, I thought about the challenges ahead. For years the field had focused on lab animals. It was easy to cure a mouse or a worm. What about humans? To study human longevity, one had to find long-lived humans and understand their genetics. Could someone tap the insight of pure science to urge a new medicine into existence? The stage was set finally to do just that. Instead of offering a reward for a Methuselah mouse gene, it was time to do so for just one person.

11 *Centenarians in the Making: 2011–2013*

In Cambodia's Sakaeo refugee camp in 1983, a young Israeli medic named Nir Barzilai stood before a line of patients waiting in the jungle mud. He wanted to improve people's lives and thought this was a place to do it. Barzilai worked fourteen hours a day to treat wounds, feed the malnourished, cure infections, and deliver babies. But then his patients returned to a disease-ridden camp. One of the aid workers had actually brought measles into the camp. The Cambodian-Vietnamese War was ending, and now other countries were closing their borders to refugees. The people he saved had nowhere to go.

Frustrated, the smallish, voluble Barzilai left for South Africa, where he treated rural villagers in the Zulu homeland. Again he saved hundreds of lives. This time he watched his patients returned to a closed society under apartheid. The philosophical medic felt defeated. Returning to the Israeli School of Technology in Haifa, he thought he had to do something different with his life. Whatever it was, it would address the politics of medicine and science.

Barzilai moved to train in endocrinology in England and the United States, focusing on diabetes. He was studying glucose metabolism in rats, first at Yale and then the Albert Einstein College of Medicine in the Bronx, thinking about the roles of environment and genetics in late-life health because untreated diabetes aged its subjects rapidly. At the time, aging research focused mainly on the rare diseases that caused sufferers to age faster than normal. If you want to help people live healthier and longer, Barzilai thought, should they not study more people who aged well?

His wife's mother, Frieda, was ninety-nine. Her father died at a hundred and two. Centenarians were the fastest growing demographic group in the world. What if he studied their genes for healthful longevity? "If I ask an audience, is there a genetic difference in human aging,

100 percent of the hands go up," the outspoken researcher explained. "The public has a 100 percent understanding that genes play a role in determining the quality and length of human life."

In the early 1990s he applied for grants to search for genes for human longevity. He was turned down. He borrowed $25,000 from his wife's family. Then he asked for his mother-in-law Frieda's DNA. His studies needed a homogeneous genetic background, and Ashkenazi Jews made just such a tightly knit group. Descended from a small number of individuals in the Middle East in the fourth century AD, Ashkenazis settled along the Rhine in medieval Germany, where they were restricted from intermarrying. They prospered, rose to positions of power, and moved across Eastern Europe.

Establishing Albert Einstein College of Medicine's Institute of Aging Research from scratch, Barzilai built it to become the biggest center for the biology of human aging in the world. He pressed the quest to find the genes that confer healthful longevity in the oldest old. From Japan to Germany to Hawaii, other research groups vied to do the same. The sample sizes were small, however, and the controls, usually siblings who did not live long, difficult to locate. The controversial research was prone to errors.

Still, by 2011, the idea had spread as several teams raced to compare the genomes of the long-lived. The competition featured Japanese Hawaiians, studied by two Canadian brothers, Bradley and Craig Willcox; Germans, studied by Friederike Flachsbart; Dutch, studied by Eline Slagboom; and other nationalities and institutions, including the New England Centenarian Study at Boston University's School of Medicine. Opponents said the studies lacked precision. There was no chance for animal longevity genes in humans because the same genes worked in different ways in humans, and human life span was not genetically determined. But the public bought in. In 2011 the Archon Genomics X Prize Foundation, popular for sponsoring a private spaceflight competition, offered a $10 million prize to researchers who could decipher the DNA codes from a hundred people older than one hundred, attracting some of the biggest names in genetic science. Barzilai wanted to win.

As the Ingeborg and Ira Leon Rennert Chair of Aging Research at Albert Einstein, he understood that science had to confront the real world of politics and policy. Barzilai lectured frequently to popular audiences, pounding home the point that aging was a unified field. You could spend billions to cure heart disease, only to die of cancer a couple months later. "That's the elephant in the room!" he said. If you want to live healthy longer, the "only MD in the bunch" insisted on exploring the underlying mechanisms of healthful longevity. To do that, study people who live healthy longer.

Spanning the world from family gatherings to business boardrooms, the centenarian gene researchers made mistakes and generated controversy. The work landed some in trouble. But they opened a new dimension to the study of longevity genes.

Turbulent Areas

Humans evolved from a primate species in Africa perhaps some five million years ago. Around fifty thousand years ago, possibly as few as a hundred and fifty individuals crossed the strait from Africa to Arabia. Our direct ancestors made only one of a number of competing humanoid species of the era, including our cousins, the Neanderthals and Cro-Magnons. Within a few millennia, we became the masters of the world. The story of our evolution is still unfolding. Perhaps longevity played a role in that history.

Certainly illness, starvation, and violence did. For most of history, the human life span was short. Most of us succumbed at an early age to famine, war, bad disease, or worse luck. A Roman's life span was thirty years, and a medieval Chinese or German's was even shorter. Not until the twentieth century could many look forward to a life past fifty.

One of the more recent developments in human evolution was that of races or tribes, which was determined partly by genetic adaptations to environment and our proclivity to mate with those closest to us. Of the human tribes, Ashkenazi Jews maintained a storied culture as they left Germany for Poland and Russia, surviving pogroms and famine to give the world such luminaries as the playwright S. Ansky, who in 1914 wrote *Dybbuk*, which dramatized the Ashkenazi myth of a dead person's spirit that can possess a living person. At their peak in 1931, Ashkenazis made an astonishing 92 percent of Jews worldwide. Accounting for five of the six million killed in the Holocaust, today they make up about 80 percent of the world's Jewish population. Supportive of science, they also made a willing test group for Barzilai's idea.

In 1998, when the American Federation for Aging Research gave him his first grant, he asked for volunteers among his Jewish acquaintances with long-lived relatives. Beginning with 384 people ranging from

95 to 110 years old, he arranged mini-family reunions with their 70- and 80-year-old "kids," about half of whom had inherited the mutation he sought. The children's spouses served as the controls. His team took family and medical histories, measured fat and body weight, did blood work, and scanned their genomes using a computer chip that read DNA like a photographer's light meter. It was an inauspicious start. Many were obese. Some drank. His oldest participant, dying just short of her 110th birthday, had smoked for 95 years. By 2008, he and his colleague Yousin Suh showed that they shared a mutation in the insulin-like growth factor gene (IGF1), one of the two human versions of the insulin receptor worm gene, which made them more sensitive than normal to insulin, and less sensitive to insulin-like growth factor. They published their discovery in the *Proceedings of the National Academy of Sciences*.

They were not alone. A half-dozen other teams raced to uncover the mechanisms of longevity in humans. Inspired by Roy Walford's book *The 120 Year Diet*, the Calgary-based graduate student Bradley Willcox moved to Japan in 1991 to study long-lived Japanese and, by 1994, as a medical student at the University of Toronto, he made his way to Okinawa. "The island was like Shangri-La," he recalled. "I wanted to learn their secrets."

His anthropologist twin brother Craig joined him, and they began a long-running collaboration with Dr. Makoto Suzuki of the Okinawa Centenarian Study, who had made the first-ever discovery of a human longevity gene. Moving to the University of Hawaii and Kuakini Medical Center in Honolulu, Bradley eventually became principal investigator of the Hawaii Lifespan Study, and together with Suzuki, the two brothers compiled the Hawaii Lifespan Study, the largest and longest-running study of centenarian longevity of some eight thousand American men of Okinawan and Japanese descent, spanning forty-seven years, as well as the Okinawa Centenarian Study, the first-ever centenarian study.

One gene "stood above anything else," a protective version of *FOXO*, the same gene that was later replicated by Barzilai's research group. If you had two copies of a single insertion on one molecule on one chromosome, you had a twofold chance of living longer; three copies, you had a threefold chance of living to a hundred. "They [the participants] hit the jackpot," said Bradley Willcox of the Okinawa Centenarian Study. With his brother and Suzuki, he wrote two books on the Okinawa diet, which seemed to turn on the *FOXO* transcription factor, instigating worker genes to make proteins that protect against microbes and oxidation, regulating responses to infection and aging.

Hot on the trail by then also was Eline Slagboom in Holland, who found the *FOXO3a* mutation critical to Dutch centenarians and moved her focus to the epigenetics, or inheritance of longevity. She saw the human version of the worm gene as a "metabolic regulator." Barzilai's group, with three others, validated this discovery in their centenarian populations, publishing the finding in *Aging Cell* in 2009. In Germany, Friederike Flachsbart, the first to replicate Willcox's discovery, was finding that a FOXO3a variation also prevented bone loss. "FOXO may provide a potential bridge between insulin signaling, free radicals, and human longevity," wrote Barzilai in 2011. In a small office at Albert Einstein, dark and stacked high with books, Yousin Suh found that FOXO3a regulated heart and brain health. It made a building superintendent controlling other workers, helping to protect and repair damaged tissues.

Their next step was to find uniquely *human* gene mechanisms of longevity. To do so, Barzilai's institute had three different efforts, one looking at DNA repair, one at mitochondrial health, and the third at human cholesteryl ester transfer protein (CETP). CETP is coded for by a lipid-modifying gene on the sixteenth chromosome that emerged in Barzilai's study even more prominently than did FOXO. Centenarians had more and larger high-density lipoproteins, the good cholesterol, whose levels were controlled in part by CETP. Mutations in the CETP gene had been associated with protection against Alzheimer's and other cognitive decline, and Merck had a drug in trial that targeted mitochondrial peptides.

Building on their mitochondrial work, Barzilai and his best friend from medical school, Pinchas Cohen, now dean at the USC School of Gerontology, developed a whole field of study based on the previously unknown mitochondrial peptides, secreted by the cell, with antidiabetes, anti-Alzheimer's, and antiatherosclerotic properties. These peptides or signaling molecules decline with age, but children of centenarians have higher levels than normal. Barzilai's mother-in-law, Frieda, had one of the highest such levels ever measured. Mitochondria are the cell's energy factories, each holding a bit of ancient DNA. Nuclear DNA contains the blueprint for the house, but mitochondrial DNA contains the wiring diagram for energizing it. They play a key role in moderating the effect of oxidative stress. Within a year, Barzilai and Cohen started their own biotechnology company to harness those antiaging properties.

It was all beginning to fit together. But just as the quest for human longevity genes seemed to be gaining respectability, one of the discoveries blew up.

Human Longevity Gene Pitfalls

By the second millennium, the concept of a gene for anything, let alone something as vague as longevity, came under assault. "Genetic determinism" became a catchphrase for criticism of a series of inflated science claims. A single gene does not create a single protein that produces a single effect. One gene makes many different proteins in many different combinations under different conditions at different times of life. The language of genetics had become a new creation myth, critics said, inflated by journalism's tendency toward apocalyptic rhetoric. We had an inadequate language for genes "pre-disposed agency," argued Ronald Green in *Babies by Design: Ethics of Genetic Choice.* Researchers found what their automated tools allowed them to find, argued others, digital correlations which had little to do with the chaos of real life.

Indeed, all but one of the human longevity studies were cross-sectional. They looked at gene expression at one moment of a person's life. Some, like Barzilai, then studied the same genes in rats and mice. The only study with longitudinal data, gene expression over time, was the Willcoxes', who could compare the Okinawan genomes with those of Hawaiians of Japanese descent. "We can even find what period of life these genes appear to be acting," Bradley Willcox said. But even theirs was a highly simplified abstraction. They struggled to see multiple genes in gene networks. "It is where we need to go. We're very interested but the tools are a little primitive," said Bradley.

On a larger scale, a gene was no longer understood as a simple four-digit sequence of information, as once taught. "Whatever a gene may be materially, it is not a unique four letter sequence of DNA," wrote Barry Barnes and John Dupre, authors of *Genomes and What to Make of Them.* The coding segments of a single gene could be separated by long stretches of noncoding DNA. These sections combined in different ways in different conditions and times. Humans were said to have about twenty-one to twenty-two thousand genes, each in two copies, one each on our twenty-three chromosomes, the other on the matching duplicate chromosome, but they needed a promoter, a switch that turned the gene on, and an enhancer, which turned on the promoter. The DNA itself did not do anything; it was rather the RNA reader which moved to the ribosome to make proteins that gave meaning to genes. The federally funded Encyclopedia of DNA Elements (ENCODE) deepened the mystery by producing a road map to the gene switches in the large sections of human junk DNA, suggesting that most diseases are not the result of changes in genes, but in their switches. The difference between the healthy and unhealthy versions of a single gene could be that between thousands of versions of rough disconnected segments.

Given all that complexity in a single cell, it was not surprising that the hugely ambitious studies of longevity in humans sometimes stumbled. In the summer and fall of 2011 several retractions appeared. First, two respected human longevity gene researchers had to retract a claim in *Science*. Boston University's Tom Perls and Paola Sebastiani had reported finding genetic markers for extremely long life, but it turned out that "the sample size was too small," said Slagboom. "You could see that from the start." More importantly, Perls and Sebastiani used a digital genomic scanning technique prone to false-positives. When peer reviewed, the mistake in their claim was discovered. It did not invalidate the work, but it did lead to a round of complaints about the drive to publish in the hypercompetitive world of longevity gene science.

Two months later, *Nature* published the counter findings by David Gems and Linda Partridge on the *sir-2.1* gene's inability to extend life significantly, as previously reported, in flies or worms. A follow-up series of reports recapped the sirtuin mistakes, with some wondering how legitimate the longevity gene field would remain, especially as related to humans. "The public pronouncements got out of hand," said one of the *Nature* coauthors, Matt Kaeberlein, rising to the field's defense. "But insulin sensitivity, rapamycin, and the energy sensor AMP-kinase are all having effects. How soon any of that will translate to the clinic is anybody's guess. We will see substantial interventions that lengthen healthful life."

Shortly thereafter, GSK/Sirtris researchers showed that they could activate SIRT1 without the fluorescent assay first used. Several European studies showed resveratrol appearing to improve the health of obese people in small groups. "The challenge is that everyone is now working on different disease indications," said Bradley Willcox. "It is going to get even more confusing before it's over."

The question remained: Which human longevity gene discovery

might yield a treatment? To answer that, researchers went to look for specific mechanisms.

The Mechanisms

"Wake up in the morning and have something to look forward to," says 104-year-old Irving Kahn, still working as an investment advisor for his $700 million private company on New York's Madison Avenue. "Work, it's the best thing for living longer," says Bayonne, New Jersey, resident Irma Daniel, 103, beaming to the camera in the Barzilai-led Longevity Project.

As the centenarian studies suggested some key genes in healthful longevity, Barzilai looked in worms, mice, and rats, and Eline Slagboom and Brad Willcox looked in human cell cultures to understand the mechanisms of their discoveries. At Einstein's Aging Clinical Research Center, Barzilai focused on two. One was a small mitochondrial protein that improved insulin sensitivity and preserved nerve cells, presenting a whole "new untapped area of aging biology," he said. Others in his lab focused on CETP, the protein potentially key to centenarians' late-life health.

In Holland, Eline Slagboom's study focused on two other kinds of proteins, heat shock and another called *14-3-3*, as critical workers controlled by the FOXO network. Genes seemed much more likely than the environment, she found, to be the causes of better heart health, and diabetes and hypertension resistance in her Dutch centenarians.

Willcox saw something similar. "Our centenarians do not die from heart disease, stroke, Alzheimer's, or cancer," he said. "Maybe FOXO enhances DNA repair or decreases free radical damage. Which is it? We don't know but we're going to find out."

The objections moved the conversation away from single gene switches to the actions of genomes, from linear cause to nonlinear, overlapping networks. In all, about six longevity mechanisms emerged from the animal and human studies—cell-stress resistance, protein maintenance, mitochondrial health, telomere lengths, cell recycling, and stem cell vigor. But to truly understand those mechanisms, to see which were damaged most or preserved be in human aging, when and where it happened, and how one such mechanism could be tapped, then the best bet was to return to the original model system which had created the field in the first place.

Worm World

Outside the window frozen Lake Michigan spread, white with winter snow. Through the Leica microscope an old friend, beautiful and haunting, edged sinuously over its golden agar beach landscape. *Memento mori*, the worm in the ointment, the creature that fed on our decay, offered the place to turn in a moment of panic. Postdoc Cindy Voisine, a calming, professional, expert voice in the lab of Northwestern University's Rick Morimoto, pointed out the neon-lit clumped proteins in a nerve, glowing in yellow. "They're beautiful," she says. "They show us exactly the moment when things start to fall apart. It's the moment when they stop reproducing."

Worms remained the beginning and end point of longevity study. When the search in humans became too slow or confusing, researchers returned to the worm. When Barzilai thought the compound heparin sulfate could affect mitochondria, he tested it in worms. When Yale's Frank Slack's small RNA workers promoted the longevity in the insulin pathway separately from the developmental pathway, he tested them in worms. When Anne Brunet wanted to test whether epigenetic changes could be passed to future generations, she tested it in worms.

At Northwestern, the youthful Morimoto focused on proteins, teaming up with Andy Dillin and Jeff Kelly to cofound Proteostasis Therapeutics in Cambridge, Massachusetts, to develop compounds to restore and preserve the ability to keep proteins functioning properly. "Longevity is about the edges," Morimoto says, trim in his gray sport jacket. "You are trying to maintain and repair an imperfect machine," he says. The twenty thousand to fifty thousand proteins in each cell fold and twist in incredibly complicated shapes in fractions of a second, in such amazing choreography as to appear almost magical. His ebullience reminded me of the lines in the bar in Kurt Vonnegut's *Cat's Cradle*, when the character asks the secret of life:

> "I forget," said Sandra.
> "Protein," the bartender declared. "They found out something about protein."
> "Yeah," said Sandra, "that's it."

Morimoto focused on protein shape as it changed in real time, using a new technology to instantaneously detect when the shape degraded.

"Of course aging's programmed," he said. "At precisely the moment the animals finish reproducing, we see proteins falling apart everywhere in the body."

At Princeton, Coleen Murphy also used worms to zone in on the very earliest moments of aging, and she too focused on reproduction. The ticking of the biological clock so many of her girlfriends worried about led her back to FOXO and mTOR signaling. "They're going to be real game-changers," she said, "because they explain a lot evolutionarily."

Other new worm discoveries yielded a variety of ideas circling back to mTOR, AMPK, and SIRT1. "That's exciting," said Coleen Murphy, who had grown tired of the field's infighting, "because it shows we have different ways to live longer."

In 2012, the sirtuins also mounted a comeback. Under the new director George Vlasuk, the sixty-employee GSK/Sirtris discovery unit published two papers explaining the ways in which sirtuin activators worked, by mechanisms that happened on different places than expected on the SIRT molecule, involving a cofactor, or helper molecule, called DCTB1. "We got it to work without the fluorescent marker," he said. Now other drug companies were competing to tweak sirtuins. "Two years ago, I would not have said that," he said. "But now we think they are."

It seemed more than one pathway worked to lengthen health. "There are different longevity programs," concluded Murphy. "We suspect you're never going to be able to live longer unless you hit at least two."

Again the field struggled to make meaning of its discoveries, and again business was interested.

New Capital

The voice on the phone in Barzilai's cramped office had a heavy Chinese accent. "My company wants to invest in longevity," the man said. "I came across your work on the web." The voice claimed he was the Houston representative of an energy company. Barzilai told him to send an e-mail. "No, my boss wants to move fast," the man said. "He can meet next Tuesday. Can you fly to Beijing Tuesday?"

"I have two hours free Tuesday. If your boss wants to talk, tell him to come here."

On Tuesday, a white Cadillac Escalade containing the president of

China's biggest private gas and oil company pulled up to Albert Einstein College of Medicine. Barzilai greeted him and began to talk. The CEO said they wanted to invest with no strings attached. Barzilai wanted control of the company with his old college roommate, director of pediatric endocrinology at UCLA, Pinchas Cohen, as cofounder. The man said Cohen and Barzilai could control 55 percent of the company, and the Chinese would take one seat on the board. "They did none of the usual background checks!" Barzilai marveled.

When I asked the man's name, Barzilai laughed. "I'll give you a hint," he said. "Mr. Lu. You can try like I did finding the right one in a Beijing phone book."

They named the company Cohbar. "Can you figure that one out?" Barzilai asked me. Based in Pacific Palisades, California, it would develop analogs of humanin, the small protein that improved insulin action and lowered blood glucose levels in rats, to attack age-related diseases, with an emphasis on Alzheimer's.

For his part, Bradley Willcox joined a company, Cardax, in Hawaii. It was targeting the upstream or controlling switch of the FOXO transcription factor with a compound and he was helping them to design the trials for heart disease.

Indeed, for all the chaotic claims and lack of success, worldwide, more people were interested in longevity gene research than ever. In 2012, *Nature* cited the manipulation of aging and the health span as one of the top four problems facing biology. In the insulin pathway, the FOXO transcription factor controlled up to five hundred other longevity genes, including antioxidants, protein folders, antimicrobials, metabolic protectors, repair and maintenance systems, and those that zeroed the clock of seed cells. The gene variant created a protective stress and pathogen-resistant state. With a reproductive and sensory modification, FOXO had created a tenfold increase in worm life. It extended the healthful life spans of mice, flies, and rhesus monkeys. Several groups of human centenarians shared a common variation in this gene. Valter Longo, having achieved his tenfold yeast life span extension, was studying a population of Ecuadorians that shared the same mutation.

The mTOR gene, in turn, was being studied for its power against cancer, triggered by the antifungal drug rapamycin. The sirtuin genes were being studied for their role in controlling metabolism and disease resistance. Resveratrol and its mimetics showed they could extend the healthful life spans of obese mice.

Roughly one hundred labs were now dedicated to understanding to longevity genes. From Italy to Spain to Canada to Holland, Japan, and China, across the world the idea had caught on. Researchers focused on epigenetics, or the changes in the expression of genes over a life time, and whether improvements could be inherited. They raced to understand the mechanisms of longevity genes found in humans.

All this happened despite the contraction of science funding around the world. Identifiable systems offered clues to regulating the quality and rate of aging through various protective mechanisms. Studies in German, Japanese, Italian, and Scandinavian centenarians had shown they shared the same variant. "The body programs the way it responds to the environment," concluded Coleen Murphy. "One of those responses is to extend normal life span."

As winter waned, I was texting my youngest sister, who had gone to check on our parents in New Jersey from her Texas horse farm. We were trying to get them to sell the home they had lived in for fifty years and move into an independent-living apartment. "I am going to lay down the law before I leave," she responded. "You are going to start packing them."

The older we become, the more our true natures emerge. The final years have a hidden purpose: the fulfillment of one's character. At the home I grew up in, the light fixture above the kitchen table had a broken plastic cover. My sister changed the bulbs and pointed out that the fixture needed to be replaced. The garage door was cracked where my father had run into it. We worried about my father driving. We thought of an appetite stimulant for my mother, who was losing weight.

The twenty-year search for longevity genes was still a young field. Arguments raged about the most basic definitions. But twenty years later, the search seemed less quixotic. Now the hunt was on for drugs.

No longer was aging thought of purely as a process of random decline, unstable and uncontrolled. "It's not what we've been told," Kenyon told a small group of biotech group leaders, bankers in black suits, along with health-care professors from Harvard Medical School, who were all gathered in the sleek Institute for New Research. She wore a blue satin blouse and yellow scarf and, at fifty-five, looked as young as some of the black-clad, would-be biotech investors in the audience.

"It's not just getting older and hanging on. Imagine you're out on a date, and you like the woman, she is pretty," she told the mostly male audience. "You ask her age, and she says 'ninety.' How would you feel?"

For the first time, Kenyon began testing compounds in human cells. As potential for some disease treatments loomed, the politics of global aging finally had to be addressed, to use Barzilai's phrase. What if we remained healthy and lived longer? What effect would humans living longer have on the planet? It was time to confront the politics and ethics of longevity.

12 *Fountains of Youth: 2013–*

My sister and I are driving our parents to a New Jersey independent-living center for a first appointment. We had been arguing about it for months, and we had finally convinced my mom to make the trip. Her doctor, a divorced man my mom kept trying to set up, said she was more removed from our conversation than ever. She rolled her eyes. She had been our father's only caregiver since his stroke ten years ago.

We live in a world that celebrates youth, but the world's fastest-growing demographic group is the "oldest old," those eighty and above. Lengthening life spans across the globe, especially in China, Japan, Korea, Taiwan, Germany, Spain, Italy, and Sweden, are dramatically altering family life and government projections. With every year, global aging will become more pronounced. A United Nations report predicts that the world's 80-year-olds will increase some 233 percent in number between 2008 and 2040. By 2040, the number of people age 65 or older should more than double to 1.3 billion. The biggest rise in the elderly's numbers is in developing countries, where one billion people over 65 are expected to be living by 2040, some 76 percent of the world total, and there is relatively little health infrastructure to serve them. Some are childless, with no one to look after them as they age. Anyone interested in longevity gene science must at some point ask: What will global aging mean for a world economy, the elderly, the planet, and all of us?

To answer these questions, a variety of nonprofit organizations have developed a wealth of reports and programs, with agendas both reasonable and extreme. The American Association of Retired Persons began in 1950 as a small nonprofit when private health insurance was unavailable to older people. It now makes a powerhouse leading positive social change for its forty million members above the age of fifty, whose poverty rate dropped from 35 percent in 1960 to 10 percent in

1995, but began to climb again in the downturn following 2008. Nonprofits like American Federation for Aging Research (AFAR), the Gerontological Society of America, the American Society on Aging, the National Council on Aging, the Grey Panthers, the Older Women's League, and others all combine to help and inform older people in this country. Similar organizations, like the International Longevity Centre, play similar roles in the United States and abroad. Business forums like Third Age Media and SeniorNet offer advice on business opportunities serving an aging economy. The World Economic Forum produces annual reports that feature some of the best international business predictions for global aging. A dozen books foresee long-range gloom or gluttony while others, like Mary Furlong's *Turning Silver into Gold: How to Profit in the New Boomer Marketplace*, offer tips for exploiting opportunities in a long-lived world.

One of the newest initiatives is the MacArthur Foundation Research Network on an Aging Society, founded in 2008 to research the best policies to serve long-lived societies, and comprised of scholars from demography, sociology, race and gender studies, business, and medicine. The world is aging faster than anyone expected, says network director Jack Rowe of Columbia University, and that challenge is inseparable from the economic, equity, energy, and environmental problems looming as some of the twenty-first century's biggest world crises.

Yet for all our concern, the longevity revolution is a fragile one. In August 2012, a study funded by the MacArthur Foundation showed that for the least-educated whites in America, average life span *decreased* by three years for men and five years for women between 1990 and 2008, a surprisingly steep drop, comparable in contemporary times only to the seven-year drop in average life span for Russians after the collapse of the Soviet Union. Life expectancy for black and Hispanic Americans of the same education level rose during those same years, the study showed, but the precipitous decline suggested that our hard-won longevity is not a given.

Meanwhile, the promise of longevity genes took hold in the popular consciousness in many developed countries. The term cropped up in magazine articles, TV shows, and books that feature the lab discoveries and phenomena like blue zones—Okinawa, Ikaria in Greece, or Nicoya in Costa Rica—where people seem to live longer, healthier lives the rest of us can imitate. Novels like Ann Patchett's *State of Wonder* and Gary Shteyngart's *Super Sad True Love Story* made the pharmaceutical race for longevity drugs a major plot conflict. The care of aging parents has been a recurring subject of contemporary German fiction. Cosmetics advertisements mimicked the rhetoric of molecular science. "Youth is in your genes!" claimed a Lancôme Génifique ad featuring an "in vitro test" that shows its "youth activating concentrate" could "boost genes activity and stimulate production of youth proteins." Avon, Hermès and Estée Lauder cashed in on the science popularity with similar claims.

Around the time my parents bought their house, the average human's life span was roughly 46.6 years. It is now 67.6. By 2050, it will be 75.5 years. The world population rose from 5 to 7 billion people in the years between 1950 and 2011, and is likely to reach 8 billion by the mid-twenty-first century. I like my parents in the house on the hill in the neighborhood we grew up. I do not want to make them move but the discussions among us, with my sisters chiming in, became so heated my mother finally said, "I'm not afraid of you." Me?

People worry that the general trend toward lengthened human lives will exacerbate generational conflict. The conflict has been given names—the clash of generations, the silver tsunami, greedy geezers, and aging Armageddon. Will the new longevity yield catastrophic conflict, pitting young against old and nation against nation, to paraphrase another book title? What will happen and how can we better shape policy to meet events to come?

Suddenly these questions feel both urgent and important. To answer them, let us run through two extreme scenarios and follow the consequences. But first, we are touring an independent-living center.

The Silver Tsunami

We arrive first in the wrong section, where several people in wheelchairs sit in the carpeted shafts of sunlight. My mother shoots me a look, until our sales representative comes to fetch us through a long hallway to the independent living lobby, with its easy chairs and a canary chirping in the corner. A gray-haired woman coos at a waiter's attention while on a poster a swimmer high-fives a young lifeguard. As we walk past the bandaged, tree-lined portico, my father smiles vaguely, as if this was happening to someone else. But he had been in advertising his entire career. He knows the pitch.

Lengthened life spans will tilt the ratio between the old, my parents, and the young, my sisters and me, sort of, who care for them. The global number of centenarians is expected to increase tenfold between 2010 and 2050, and the number of older poor, the majority of them women, to go from 342 million people today to 1.2 billion by that same year. We may celebrate youth, but we live in the Age of Aging.

What does that change mean? First, it shifts the economic balance between young and old. The Social Security Administration projects the rate of life span increase to slow, but competing projections from demographers like S. Jay Olshansky and others have the balance toward the elderly tilting faster for certain subgroups of the population, with average life spans lengthening a year every year through 2050. The Medicare and Social Security deficits would, if policies remain unmodified, shoot higher than expected, bankrupting Social Security in twenty-five years. Global population would increase and the percentage of elderly in societies would grow. Right now approximately four Americans work for every retired person. By 2050 it is predicted to be two, unless more and more of us work past the ages of sixty-five or sixty-eight. By 2045, people older than sixty will outnumber those under fifteen for the first time. Japan's and Russia's workforces are expected to shrink by nearly one-third by 2050 and Germany's by one-fifth. The number of older people, some of them depending on the young, will rise just as the number of young people will fall.

Second, women will be most affected, comprising the majority of the older population. Today, women outlive men by an average of six years. Eighty percent of the world's centenarians are women. Though both sexes today live longer than ever, the gender longevity gap remains. That means that eight times out of ten when we speak of the oldest old, we are speaking of women without male partners.

Third, this aging trend promises to accelerate for most groups. "Forthcoming advances in the biomedical sciences will lead to interventions that slow the rate of biological aging and have a systemic dampening effect," Olshansky recently wrote, "on all fatal and disabling diseases simultaneously." By 2050, the average human life span increase could be as much as thirty years, suggests Shripad Tuljapurkar, demographic programs director at Stanford's Center for the Demography, Economics and Health of Aging. If we are living longer with incremental improvement in medical care, then any drug or molecular genetic intervention would tilt the balance even more.

This should be considered a terrific trend. A healthy later life maximizes your opportunity for a fulfilling human experience. It is also good for a national economy. Harvard professor David Bloom, editor of a World Economic Forum book analyzing aging's economic effects, argues that improving late-life health improves productivity, strengthens education (with more time and money to invest in schooling), increases savings, spreads gender and ethnic equality and enhances opportunities for young people, coupled with an overall reduction of poverty. By comparing two countries, identical in all respects, except one has a five-year advantage in life expectancy, Bloom showed that income per capita in the healthier country will grow 0.3 to 0.5 percent per year faster than in its less-healthy counterpart. Healthful longevity is not only an effect of a better economy, Bloom argues, but actually a cause.

Others disagree. A twenty-one-year-old looking for a job may be told they are being held back by slowing retirements. A sixty-one-year-old worrying about keeping a job a few more years might not consider her long life a good thing.

Regardless, whatever may happen with global aging, governments have time to adjust. The problem is that no one is sure exactly *what* to adjust, either in the developed world or the booming African and Arab countries of the developing world. Global aging will require many changes in the way we think. To understand how let us explore some qualities of two extreme scenarios and follow the potential lessons.

Aging Battlefield

In Minneapolis, a fifty-seven-year-old health insurance auditor named Roxanne Aune sometimes feels she cannot take it anymore. She works full-time while caring at home for her fifty-nine-year-old husband who suffers from Alzheimer's. She feels passed over for important work projects but dares not mention to her employer how difficult her life is. "I think discrimination, subtle and unprovable, would get worse," she told the *AARP Bulletin* in October 2011.

Global aging means that the challenge of caring for the elderly will impact Social Security and Medicare in the United States, feed the euro, health care, and ecological crises abroad, but also cost workers and companies in lowered productivity and heightened stress. An expanding elderly population already stresses government and family budgets and personal lives. 16 percent of the national gross domestic product goes to health care in the United States, but some 25 percent of total American health spending goes to the elderly. In Asia and Europe, the median age is projected to rise from thirty-six today to fifty-two by the middle of this century. In Spain, when the Socialist government tried to raise the retirement age from sixty-five to sixty-seven and freeze pensions, riots wracked Madrid. The same thing happened in Paris when the retirement age raised from age sixty to sixty-two. In China, according to Ted Fishman in *Shock of Grey: How the Aging of the World Population Will Pit Boss Against Worker, Generation against Generation*, one worker will be supporting six people on average within a decade. While many of the elderly are active workers, Fishman claims, others will be dismissed from jobs simply for being older and overcompensated. In a global market, he writes, "age discrimination pays."

In short, the doomsday scenario suggests that global aging raises many of the pressure points of the twenty-first-century world—economic, ecological, and political. It may force developed countries to choose between funding entitlements or investing in education and infrastructure. Developing countries are even less prepared for the demographic shift, with fewer resources and less formal training in geriatric health care to support older people's needs. If the elderly have children, they are likely to be cared for. But if the elderly in a poor country are childless, they fall upon a government's meager care. This prospect can in turn drive up birth rates, deepening the world poverty cycle.

Our past success generated these problems. A century ago we would not have lived long enough to worry about care or health or retirement in our nineties. We have lived through a biomedical miracle joining new drugs and technologies with improved delivery systems and rising wealth. We live longer and, for the most part, better than any previous generation. By 2050, the number of Americans over the age of fifty will be 20 percent of the population, up from 8 percent in 1950. If they are well-to-do, it is a good time to be aging.

But it is a bad time to be poor and aging. In a global economic crisis, much of the shock falls upon those on the fringes, like the elderly. The alarming recent drop by four years in the average life span for American whites lacking a high-school diploma seemed to have several causes, including rising obesity, job stress, and smoking rates and declining access to good health care. Nationwide, as we pass the age of sixty-five, the probability of being poor increases, as does the probability of suffering from mental illness and depression. Medical bankruptcies today affect those sixty-five or older more than any other demographic group, while the cost of prescription drugs is rising. The gap between rich and poor is growing fastest among the elderly, of whom some 80 percent suffer from some form of chronic health condition. Today, nursing home care takes 6 percent of the gross domestic product in the United States. That percentage will surely rise.

Caregivers for the elderly are particularly hard hit. More than 44 million people are caring for an aging adult in America; around 60 percent of these are women. As the elderly face debilitating illnesses, their children—when they are lucky enough to have children—face loss of work time and the expense and stress of caring for aging parents. When you become a caregiver, there is often little advance notice. It can put a person's job security at risk. A 2006 MetLife and National Alliance for Caregiving study suggests that businesses lose some $33.6 billion a year in productivity for caregiver employees.

Indeed, a whole industry of businesses has sprung up to serve the needs of caregivers in the sandwich generation, boomers with children who themselves are taking care of aging parents. It could be that caring for elders will make the new midlife crisis. Insurance for personal care, reverse mortgages, and corporate eldercare programs all make for growth industries.

The biggest problems in caregiving would likely face aging women. The journalist Susan Jacoby quit work as a reporter to care for her husband when he was stricken with Alzheimer's. There is a difference, Jacoby notes, between being well-to-do and old or poor and old. The experience varies widely by ethnic group. Focusing on the plights of aging women, sometimes alone, poor or caring for another, she writes:

> Here's what one cannot do . . . complain about health problems to anyone younger; weep openly for a friend or lover who has been dead more than a month or two; admit to depression or loneliness; express nostalgia for the past (personal or historical); or voice any fear of future dependency.

The irony was that in her early magazine career, Jacoby penned a number of the utopian "this is the new aging" self-help articles that make up a freelance writer's staple income.

There is some reassurance in knowing there is nothing new in this crisis scenario. The first contemporary aging scare came in the 1980s as the first wave of baby boomers reached sixty, then an advanced age. In 1980, a *Time* cover featured "Greys on the Go," while the *New Republic* coined the term "greedy geezers." University of Chicago researcher Alan Greenspan authored a nonprofit report, *Aging America: Who Will Shoulder the Burden?* MIT economist Lester Thurow and industry executive Peter Peterson each made whole careers out of penning bestsellers detailing the coming aging disaster. As it turned out, the increase in life expectancy in the 1970s and 1980s added some $57 trillion in contribution to gross domestic product as people worked longer, accumulated more wealth and savings and contributed for a longer period to Social Security, all of which brings us to the next scenario of an aging world, heaven.

Aging as Paradise

Around 1981, S. Jay Olshansky, then a sociology doctoral student, was sitting in his jeans and T-shirt in a University of Chicago seminar. The Detroit-raised contrarian was something of a rebel who thought things did not have to be the way they were. When his teacher, Bernice Neugarten, said, "There are the young old, and the old old," Olshansky's hand shot up.

"How can you say that?" he said. "All old people are all alike."

Neugarten stared at him. "Jay," she said, finally. "I have an assignment for you. I'm working on a project to determine whether the government should get into the business of slowing aging. I want to know what the demographic consequences of success would be. You go to the library and find me the answer."

He went. He got hooked on the simple question of how long people can live. Reading the first demographers of aging, Olshansky was plunged into a conflict between two groups fighting over the terms of the debate. One group, evolutionary biologists following Darwin, ignored the biology of human body design. The other group, population experts, followed the prognostications of Benjamin Gompertz, who predicted in the 1820s an exponential increase in death rates as we age. Olshansky became so entranced he switched his focus to longevity and helped invent the modern version of the biodemography of aging, the study of the underlying biological forces that influence the duration of life. His PhD dissertation proposed a new way of calculating life expectancy still in use by some insurance companies. In 2000, seeing the power of the new lab discoveries, he won a National Institute on Aging grant to learn the molecular genetics of longevity. He coauthored a book, *The Quest for Immortality*; participated in boards and panels; appeared on programs and in a *Scientific American* cover story; and gave talks at the Davos, TED, and Google visionary-type science conferences; all while starting his own consulting company to analyze the effects of global aging.

In his cinderblock office on a late fall afternoon in Chicago, the short, balding demographer wore a black golf shirt, jeans, black sneakers, and eyeglasses. He rubbed his face, which was trimmed by a gray beard. On the powder-blue wall hung a three-foot Indian tapestry flanked by the poster for the 40th Nobel Conference, held in 2003, on the biology of aging. A dusty oversized martini glass and a Chinese figure of longevity, an old man and an apple, decorated the bookshelf. Back from Abu Dhabi, Olshansky had conference talks upcoming in Seattle, Washington, Haifa, and Beijing, as he sought to correct the negative vision of global aging. He had read about an aging workshop at a Midwestern medical school and was upset at the portrayal of elderly life. "They had their medical students put on braces and weights to simulate the infirmities of old age," he told me. "I said: 'You're sending the wrong message here!'"

Olshansky argued for the idea of the longevity dividend, that living longer, healthier lives will provide a boost to the world economy much like the peace dividend that followed the fall of Communism. "Longer life does not lead to overpopulation," he pointed out, explaining that the longer people live, the fewer children they have, leading to more opportunities for women to go to school and get better jobs. "In fact, healthy old people power the economy," he says. "They go on trips, buy things; actually they are the biggest consumers of health care. Older people are this huge untapped resource."

At the moment, extremists on both sides owned the longevity debate, he complained. Take retirement. Economists simply do not agree whether compelling people to work longer will save money or detract from opportunity for the young. The young and old do not compete for the same jobs. Along with his MacArthur Research Network on Aging colleagues, Harvard's Lisa Berkman and Columbia's John Rowe, Olshansky feels the idea that delayed retirement will reduce the number of jobs for the young is a "trap." People should be free to decide when to retire. Some might choose to do so early. "If you're in a manual trade, or have a lower life expectancy, or you're unhappy in your job, you should be able to retire early," Olshansky said. "For others like my teacher, Bernice Neugarten, it was a crime she was forced into retirement at the height of her career." There should be more job sharing, better leave policies, and individual, phased retirement options.

If this could be the best time in history to be getting old, as Olshansky argues, he is echoing researchers like technology analyst at the Pacific Research Institute Sonia Arrison, author of *100 Plus: How the Coming Age of Longevity Will Change Everything From Careers and Relationships to Family and Faith*. The last eighty years produced the first social net for the elderly. Social Security and Medicare were created in the 1930s as part of the New Deal, and comparable programs appeared even earlier in Europe. German chancellor Otto von Bismarck is credited with instituting the first social insurance program for the elderly in 1889. Today, the Internet, social media, and improved awareness of lifestyle possibilities are improving the experience of growing old. Giving people better access to education and benefits could make the aging generation a catalyst for social change. It is a time of potential social activism and engagement for older people.

As for overpopulation, at first some demographers thought life extension would burden the planet, as Stanford's Shripad Tuljapurkar remarked at the American Association for the Advancement of Science conference in 2006. But it turned out that lengthened life span is not a significant cause of overpopulation, as feared. At the University of Chicago, a husband-and-wife team, Leonid Gavrilov and Natalia Gavrilova, devised an algorithm that uncovered the exact opposite conclusion to that of Thomas Malthus, who did not account for the productivity explosion from the industrial and information revolutions. Current health improvements have been accompanied by an astonishing, beneficial fall in fertility. The transition from a fertility rate of five children per woman to two took a hundred and thirty years in Britain, but just twenty years, from 1965 to 1985, in South Korea. In Iran, the fertility rate fell from seven in 1984 to just shy of two in 2006. This falling fertility means better security for billions of vulnerable people, particularly women. The countries with the longest life spans have the least issue with overpopulation, which is rather a product of poverty, inequality and disease.

We have a model, the heaven scenario suggests, the previous longevity revolution from 1900 to 2000 when average life span increased by some thirty years. During that time, world gross domestic product increased nineteenfold, by some $39 trillion. Rather than starvation due to population growth, one result was widespread obesity. The current economic crisis is not a crisis of agricultural scarcity, but rather of overabundance. Everything depends, to paraphrase William Carlos Williams, on our health as we age. This is where the biology of aging had taken us, and it makes our final stop.

Demography Is Not Destiny

Government planners can either ignore global aging and then scramble individually, or cooperate to adjust institutions and policies together. Healthful aging can be better for the bottom line, families, and individuals. If we can address the diseases of aging in a unified way, we can decrease elder care costs. For most of us, about 90 percent of our total lifelong medical costs will be racked up in our last year of life. The key to the future, then, is a compression of morbidity, remaining healthy until we die, as centenarians do. A new World Economic Forum anthology, *Global Population Aging: Peril or Promise*, offers a few common-sense policy insights.

To get there, first, some simple and seemingly obvious government-supported lifestyle changes can yield tremendous benefits. The American government's antismoking campaign reduced the rate of smoking from 40 percent in the 1950s to 32 percent in the 1960s and 1970s to 19 percent in 2007, hugely reducing deaths from lung cancer. It was a success, although the numbers recently have crept back up to 21 percent. Perhaps an understanding of diet and recreation for young can do the same for diabetes and obesity. There needs to be better nutrition at an early age. In children worldwide, even a small vitamin deficiency can damage DNA as much as radiation. In the 1930s, the United States started adding Vitamin D to milk, virtually negating rickets in northern states and improving bone health in the elderly. The easing of hunger for pregnant women or infections in very young children could prevent late-life inflammatory diseases like diabetes and hardening of the arteries.

Sleeping longer, improving diets, moderate exercise thirty minutes a day, all can play a role in aging with health. The government can help. Already the new Medicare prescription plan, focusing more on prevention than acute care, is paying dividends as more seniors are getting access to medications before their illnesses became acute.

A number of reports, from the US Senate, the MacArthur Foundation Longevity Research Network, AARP, Kiplinger, and others, shared a similar prognosis for fixing Social Security, one of the most successful government programs in history. It is slowly going broke under the formula devised at a time when life expectancy was twenty years shorter than it is now. Some of the relatively straightforward steps to maintaining solvency include a slight increase in contributions, a rise potentially in the retirement age by two years, and an increase in the maximum amount of contributions.

Other scholars have explored similar adjustments in other government programs and social attitudes that would help accommodate the new longevity. For example, research conducted by University of Chicago's John Cacioppo showed that keeping elderly people in the community will extend their health spans and limit obesity. Research by Stanford's Laura Carstenson, founding director of the university's Center on Longevity, suggested that memory is improved by motivation to improve, provided by an attachment to an emotional community. Research by German economist Axel Boersch-Supan and sociologist Martin Kohli demonstrated that sharing workloads among the oldest and youngest professionals improves productivity. Older workers possess invaluable experience.

If Social Security can manage in the United States with minor adjustments, for Medicaid, the crisis is bound up in the larger issue of health-care spending, which in 2011 actually showed its smallest annual increase in a decade.

Supporters of the molecular genetics of longevity proclaim these numbers as reasons for more funding for basic research. Current medical research is somewhat separated into disease empires, some critics claim, hindering science's ability to apply the longevity gene discoveries thus far. Research into the biology of aging to improve the overall quality of late-life health is more necessary now than ever, argued Robert Arking at the end of his book *The Biology of Aging*.

There is a battle for the right to speak on these ideas. For years, the dialogue was dominated by some of the more visionary speakers at the edges of science. The Cambridge engineer Aubrey de Grey, for instance, argued we should not accept chronic disease at all, and that ours will be the last generation to experience the disabilities of age. The entrepreneur engineer Ray Kurzweil proclaimed that were in a historical moment of new "singularity," a shift in the fundament of human expectation. Both figures spoke at conferences and have become the subjects of books, while producing books of their own. Both are fascinating figures with a larger audience than the corresponding professional scholars in the marketplace of ideas shaping our time.

There is a battle, yet some scholars have eschewed the combat. When Jay Olshansky argued against the possibility of a maximal life span in the journal *Science*, it brought him head to head with opposing demographer James Vaupel at the Odense University in Denmark. Vaupel began avoiding him at conferences to the extent of not attending if Olshansky was scheduled to appear. "That's not science!" Olshansky said. "Science is all about free discussion and disagreement! That's the currency of ideas. Jim doesn't feel that way. He thinks humans can live to a hundred and fifty and I am sure they can't." Smiling in his cinderblock office, Olshansky leaned into my digital recorder. "You're wrong, Jim!" he said. "Do you hear me?"

Eight months later when the two scholars finally did meet at a confidential European insurance company conference to analyze the economic effect of global aging, they resumed their disagreement.

In the Moment

Every research group has an angle and agenda. Whether it is the prolongevitists or the antielderly, each speaker and thinker pushed a plan that served, in part, their self-interest. That is the marketplace of ideas and, in matters of dispute, the science of the new biology of aging resembled the play of political theory. In such a unique research field as the molecular genetics of longevity, we may well learn to treat researchers like politicians. Science, one researcher once told me, does not belong to the scientists.

Science should provide a right answer. The right answer to the question of global aging is this: living healthier, longer is the signal accomplishment, along with equality and the rising wealth and knowledge of human civilization. Any research that can further human health span, the new word of the current moment, is to be supported.

Improved social policy, greater social equality, and some health-care overhaul are necessary to address the issues of global aging. Some Asian and European countries can provide a model, according to the World Economic Forum. Shared work, flexible hours, and scheduling will play a part of the job age revolution. Integrated communities could do the same for a potential housing age revolution. It should be remembered that older generation offers services that can assist and transform society and that a part of the solution is an attitude change for all age groups. The scholarly field of age studies is to refocus attitudes toward the treatment of age in the media and literature, to do for aging what women's studies did for women or African American studies did for African Americans. Scholars like the University of Washington's Kathleen Woodward insist that people can find new opportunity and passion in their later years. "If older people feel they have little to contribute," Woodward writes, "they fill that expectation. The key is to change the expectation, in which case old age may become a time of greater passions and commitment to help than even in youth."

We experience many direct benefits from a healthier older population. Volunteerism is big business, worth $272 billion in the United States in 2009, some 2.5 percent of gross domestic product, and volunteerism became the subject of a joint Harvard School of Public Health, Metropolitan Life, and NIA advertising campaign featuring John Glenn, Quincy Jones, and Martin Sheen. Listen to a master class conducted by a great artist, like violinist Yehudi Menuhin, seventy-seven, and the value of wisdom in the arts is clear. The same can be said for business. The value of elder contributions to society is so great in fact, in 2007, the rock musician Peter Gabriel joined with Virgin Airlines CEO Richard Branson to found the Elders, an international consortium including Nelson Mandela, Desmond Tutu, Gro Harlem Brundtland, Norway's first female prime minister and a champion of sustainable development, among others, to advise nations on the global issues. When U.S. Airways flight 1549 lost both engines in 2009, its fifty-seven-year-old captain Chelsey Sullenberger had the experience to ditch safely in the Hudson River. Sixty-eight-year-old Elders member Graca Michel, widow of Mozambique's former president, had the experience to defend African children of armed conflict.

In America, Social Security and Medicare remain some of the most successful government programs ever, but in 2013 each required a tune-up to serve an aging population, quite reasonably, as each program had been designed for a less healthy country. To keep Social Security solvent, planners may consider raising the cap (upper end incomes have risen dramatically since the programs began), raising the payroll tax rate on the highest incomes, or modifying the cost-of-living adjustment. For Medicare, similar adjustments may be considered—raising the eligibility age to sixty-seven, or raising the Medicare payroll tax or premiums on the wealthiest, for whom Medicare is a relatively small part of their retirement benefits. One thing is certain: everyone must be careful whenever a proposal is put forth to "save" either program.

In the labs, meanwhile, researchers continued to make new discoveries even as they were invited to join policy debates. Valter Longo was invited to explain to congressional committee members his discovery that fasting during chemotherapy improved patients' resistance to radiation damage. Researchers in Italy found a longevity compound in semen. Others focused on omega-3 or coenzyme Q. Could aging decline be something as simple as losing the ability to respond to environmental stress? One key area of research remained mitochondria, the cell's energy factories. As regulators of the rate and efficiency of oxidation, they play a key role in moderating oxidative stress. As we age, our mitochondria slow down, leak, and become inefficient. They each have a bit of ancient DNA, which is highly subject to mutation as we age. The key figure in the study of mitochondria and aging was the Children's Hospital of Philadelphia's Douglas Wallace, who noted that the importance of the cell's energy factories had been too often overlooked by longevity researchers "due in part to the dominance of the anatomical and Mendelian paradigms in Western medicine."

The global aging crisis is bound up in the global energy and environmental crises, and solutions will be slow, sometimes painful, but interlinked and also potentially exciting. As we drove my parents back up the Hudson River to lunch in Peekskill, they told my sister and me they were not ready to move. Nor were we ready to make them. I wanted them to live forever. But we were worried. The question was, what could the newest research show?

13 *Reimagining Age*

At the new Marin estate of Cynthia Kenyon, I am hiking with her in the mountains. "I got married, you know," she says, to Jasper Rine, geneticist and biotech entrepreneur.

"My grandma's advice when I got married was 'be nice,'" I tell her.

"My father said the last three years of his marriage were the happiest in his life," she says. We walk along the rugged trail flanked by George Lucas's estate. "I think your book could be majestic," she says. "When are you ever going to finish?"

Most science proceeds quietly. Problems are tackled by one or a few laboratories and the results published in a journal. The effort to find longevity genes began like that, but competition and public notice changed the research into an extraordinary scientific venture, featuring big money, big personalities, and a race with enormous potential consequences. Moving in twenty years from utter obscurity and disrepute to the pinnacle of investment and media hype, it experienced an influx of funding and changing motives that altered the discussion of ideas. Analysis became clouded by personal opinion and financial interest. *Oprah* and *60 Minutes* became the soundstages of science. The show, and not the conference, was the venue at which much of the molecular genetics of aging was performed and evaluated. Motivations that began with the simple curiosity to use new tools to ask old questions inevitably became tangled as money poured in.

Science is never perfect. It bears little resemblance to the textbook progression taught in schools. It is as chaotic and flawed as the people who pursue it. When researchers discovered that biological systems could prolong their survival longer than previously thought, the potential was clear. Mammals in the most inhospitable environments could evolve amazing stress resistance and protein preservation properties to ensure their survival. Through imagination, mistakes, and grueling work, scientists uncovered some of the mechanisms of extended health. Investors swooped in.

With drug companies funding early clinical trials, new technology tracking specific proteins and enzymes, and experiments charting changes in gene form over a lifetime, the quest turned to a new unifying theory of longevity. One emerging model was that the genes behaved a bit like investors, picking and choosing the traits that would maximize their return, yet susceptible themselves to mistakes and faulty communication. As new researchers entered the field, the lab battles resumed and the feeling grew that something important was bound to come through. The biggest discovery could be something accidental or overlooked. Research sought to move from lab to clinic.

That move required a tricky balancing act. Most every biomedical researcher wants to discover something beautiful and fundamental that will improve human lives. That was exactly what scientists thought they had done. But to move to the clinic requires massive manpower and money to investigate toxicities, disease applications, and conduct clinical trials. To get from lab to clinic required that companies patent their founders' discoveries, sign up other researchers to exclusive cooperative agreements and partner with pharmaceutical corporations to develop drugs. Many of the same researchers who signed cooperative agreements with companies served as editors and reviewers on the major journals and funding agencies of their fields. These commercial ties could well have altered the public perception of the science: if a young researcher submitted an article or grant proposal contradicting a company's professed platform, reviewers with commercial ties might not have been motivated to accept their proposals. The money affected individual interactions and the free flow of information.

There was never a science like this for its sudden media and money infusion, neither cancer, nor polio, nor the newer fields of RNA interference, monoclonal antibodies, or stem cells—nothing that resonated so immediately to the wider public while drawing on incredibly complicated, completely disparate findings in many different molecular signaling pathways of almost of infinite complexity. By 2012, companies and labs raced for compounds to tweak the insulin, sirtuin, mTOR (the mammalian form of the protein targeted by the Easter Island compound), and other pathways, including those known, like rapamycin, metformin, and resveratrol or its imitations, and those unknown or kept secret. A few studied the aging of animals in the wild for clues. With new tools, lab research turned to the minutest mechanisms of cells as they aged, including protein decay, organelle recycling, response to stress, telomere lengths, mitochondrial efficiency, and the gene packaging and expression patterns of youth spread over a lifetime.

If the story of longevity genes made a parable of science in our era, it is time to assess the results and handicap the future. To understand where we are going, we need to take a brief look at the past and present of a singular science story.

Famine and Flood

The human body is an amazing feat of engineering. If you unspool the DNA inside our trillions of cells, it would stretch from here to the sun and back, many times. We are each a biomedical gold mine, with a remarkable brain, an immune system honed over ten million years of evolution, and natural abilities to withstand famine and flood, while making and bearing children and carrying on our lives in the midst of chaos. It is a gift from our ancestors, complete with its own nutrient-sensing signals that may calibrate our healthful lives.

We have lived through a revolution in our understanding of these default signals. Twenty years ago, the idea that the decay of aging could be slowed was heresy. Now we know that a few simple changes, along with potential understanding of molecular genetics, may slow that relentless tide. These changes include daily exercise, reasonable nutrition, and engagement with others and with life.

Today's world of junk calories and lack of exercise disrupts the insulin signaling within the body, increasing glucose levels and uptake. We need to reduce that signal, to return to a world of reasonably controlled diets. Amazingly, some element out of four major interconnected signaling pathways may in some way help us do so.

The first and most widely important is insulin and insulin-like growth factor, borne out by research in a dozen organisms and almost as many human studies. Glucose is the fuel and insulin the distributor. Raising the gas pedal extends the life of the car. It does so by regulating proper protein maintenance and oxidant, microbial, and stress resistance. The second pathway could well be the metabolic silent information regulators, the sirtuins, shown in some cases to have an effect on life span in yeast and in numerous studies of obese mammals, along with promising human studies. This regulator of ordered gene expression controls the production of energy and response to illness and to exercise. The third was the gene called mTOR, target of the compound found on Easter Island, a major nutrient sensor most active if you have a lot to eat. mTOR is part of ancient signaling pathway that overlaps with insulin and insulin-like growth factor signaling. The fourth was the energy sensor AMP-kinase, the workhorse. Each path overlaps with the other. Into each or these molecular pathways, and others, labs poured their resources in the millennium's second decade.

To name a few such labs, at the University of Michigan Richard Miller studied cell-stress resistance as the prime driver of healthy aging. At Northwestern University, Rick Morimoto focused on protein quality. Princeton's Coleen Murphy focused on reproductive health. Rutgers's Monica Driscoll studied compounds to activate *FOXO* genes. Numerous researchers at the University of Texas studied the effects of the immunosuppressant drug rapamycin. At the Mayo Clinic, the focus was on ridding the human body of senescent cells. Drake University's James Christiansen studied the telomeres of long-lived wild turtles.

Many turned to the epigenetics of aging, the ways in which the ordered expression of genes over time is controlled by marks on their casing. Stanford's Anne Brunet and her graduate student Eric Greer have shown that a change could be inherited for four generations of worms. She and many others raced to uncover the changes in specific genes that drove healthful longevity, and their environmental triggers. "Food, stress, hunger, radiation," Brunet said, ticking off the possible sources. "It could be illness or pollution." It could be that love, as nurtured, long-lived rat pups suggested.

The study of stem cells in aging offered clues to what youthful replenishment might look like. Tom Rando investigated stem cells' roles in rejuvenation of muscle tissue and, in collaboration with his colleague Tony Wiss-Coray, turned to the brain. Anne Brunet also showed that the transcriber FOXO3a preserved stem cells that repair our brain neurons as we age, while researchers at San Diego State used stem cells to rejuvenate aging hearts. Such research turned to the epigenetic DNA marks that dictated to cells the exact organs they might generate.

Caloric restriction, in the meantime, took a hit. By the spring of 2012, the two long-term studies of rhesus monkeys revealed conflicting results. The Madison, Wisconsin, monkeys showed significant improvement in health and longevity due to caloric restriction; another group at the National Institute on Aging lab did not. The NIA study animals did, however, have a diet much reduced in sucrose, closely resembling a diet in the wild, and they did have a slew of signs of better health than those on regular diets. Some of the NIA male monkeys reached a record-tying forty years of age. The message seemed of to be that by staying away from processed sugars, one might conceivably eat almost as much as one wants and still gain many of the health benefits of limiting calories.

Others stayed with the old standbys, like caloric restriction and mitochondria, heroes of the free radical theory of aging. Rozalyn Anderson at University of Wisconsin–Madison studied caloric restriction in mice and monkeys. Johan Auwerx in Switzerland focused on resveratrol and metabolism.

The debate had moved beyond principle to specific mechanisms of longevity. "When I entered the field, the question was whether tweaking genes affects aging," said Andy Dillin, who moved to the University of California at Berkeley. "The question now is which genes most affect longevity." To answer that question, a brief look at the past would point toward trends for the future.

A Brief History

In the beginning, the early application of evolutionary theory to the problem of life span tended to block a pathway to truth. Researchers could see some species lived a lot longer than people, like some turtles, some species of fish, and redwood trees, but did not seek to explore the significance. "Most people look at that and think, oh, isn't that interesting. But then they think I don't want to be a turtle or a fish or a redwood tree. They didn't have a role model," observed Cynthia Kenyon. "They did not ask why similar animals might have such different life spans."

Tom Johnson's breakthrough discovery was exactly what Mendel did in the nineteenth century, showing that one gene was responsible for a trait. The shock was, the trait was life span. "The molecular geneticists did not believe it. The evolutionary people did not believe it. Twenty-five years ago, they said there was no chance at all for humans. They were dead wrong. We came to see those genes working in mice in exactly the same way as in worms. All those genes had human homologues. The insulin, IGF-1 pathway along with other genes, really now make a network," concluded Johnson.

Genetic study in the worm *C. elegans* began a revolution that allowed the study of cause and effect in longevity with much greater analytical precision than before. Suddenly researchers had the molecular tools to dissect the processes of decay, and grab a foothold on feedback systems a billion years in the making. First came the longevity mutants in worms and yeast, coupled with gene cloning and analysis to identify the mutated genes and determine what they did. When it was seen the worm gene coded for an insulin and growth factor receptor, just as in humans, the research was ignited. Then researchers turned to the new technique of RNA interference to conduct larger, unbiased screens of all the longevity genes in mice and flies. The same, and new, circuitry that determined fitness could be seen in these organisms. If that was the case, then there was a good shot at finding them in humans.

Since the insights were so new and varied, "every time they turned around it seemed they made a new discovery," recalled Kenyon. The capabilities of wild animals to live longer finally came to the fore. The study of the manner in which genes turned on and off had become the major problem of molecular biology. Now one could apply those insights to the most intimate, biggest question of all—how long would we live.

In 1996 in Carbondale, Illinois, Andrzéj Bartke showed an insulin- and growth hormone–deficient mouse could live nearly twice as long as normal. Coupled with that, the Ruvkun and Kenyon analysis of *daf-16* (Sweet Sixteen) in 1997 vaulted the research from an obscure worm into human aging. The news media noticed. NBC News profiled the Ruvkun lab as numerous other labs around the world turned to the components of the signaling pathway in worms, flies, and mice. The scientists filled in the missing pieces from other insulin pathway discoveries in several labs from 1997 to 2000, traced through a series of genes and transcription factors. A metabolic pathway determined life span, just as everyone had thought. Think of the body as a building. Its thermostat is a nutrient sensor. Its intercom communicator is a hormone. The building superintendent is a transcription factor. The human version of the transcription factor, FOXO, hires the painters, sweeps the floors, repairs the plumbing, and maintains the building, turning on worker genes that protect us against microbes and oxidative stress.

After the human genome sequence appeared in 2000 came a race to monetize the genome. Some of the longevity gene researchers pitched a synecdoche of transformation. Just as technologies changed over time, so too did the associated ideologies. The years 2004 to 2008 made a critical moment when the goal changed from that of pure science to one of transforming society, and a plant compound from red wine appeared to offer an entire platform for making drugs. It seemed like the perfect story. There was a model of such innovation in universities being turned to companies, like Biogen, Genzyme, and Vertex, which in those same years brought to market important drugs for cancer, hepatitis, cystic fibrosis, and other diseases. Biotech fulfilled the promise of the information revolution, and some of the moguls of the information age jumped into it.

With intense media interest, there followed an explosion of competing agendas, investments, and egos. The science moved at high speed only to fall into the traps of slow-moving reality. Some investors sought a quick flip, to start and take companies public before their science could be exposed. The research was difficult and costly. To get money you had to entice the public. The company Elixir came too early and promised too much. It did not generate quite enough attention. Sirtris by contrast was aggrandizing and committed to its discoveries. It simplified the results and for a period of time the public pronouncement outpaced the science. In the end, the appeal of the longevity genetics research made for a kind of power. But the combat over public perception and the ideals of business and profit pushed science much faster than if the combat had never taken place.

There were mistakes and the science was difficult. The compound resveratrol extended life span significantly in some yeast and fly studies but not in others, and in some studies of obese mice but not others. It appeared it might work best in the right dose, timing, and tissues. Overexpressing the *SIRT1* gene did not lengthen life in three mouse studies, but improved health and might work if done in the right tissues at the right time. The sirtuin genes played a role in the life extension conferred by caloric restriction, but the effect was complicated and dose-dependent. The claim of activators one thousand times as powerful as resveratrol was misleading, but other second-generation sirtuin activators were under intensive study as treatments for metabolic and other diseases.

Initially, the contrary evidence was relegated to conference presentations and a web-based journal, followed by larger but still specialized publications, picked up by a pharmaceutical blog and ultimately in well-researched reviews at the end of 2011 in *Science* and *Nature*. Yet the original resveratrol, sirtuin, and caloric restriction claim, flawed though it was, spawned an important field of study. Sirtuins made a leading edge of the new epigenetics of aging research and resveratrol did seem to increase the power and number of mitochondria, the fitness of obese animals, and, perhaps, humans. If one looked downstream at genes like *SIRT3*, as the University of Wisconsin's Tomas Prolla was doing, then one saw a connection to caloric restriction." The detail is critical," observed Wisconsin's Rozalyn Anderson.

In May 2012, resveratrol made the cover of *Cell Metabolism* as David Sinclair's lab and an international team demonstrated that genetically engineered mice fed a high-fat diet and resveratrol experienced improvement in their health mainly when the *SIRT1* gene was present, suggesting the compound activated the gene. The Elsevier (*Cell Press*) press release seized the finding to proclaim, "[This] resolves the controversy on the life-extending compound." However, it was unclear what was happening metabolically with the lab-created mice, leading the science website Embargo Watch to call it a "truly appalling press release." The controversy continued.

The years of the longevity gene boom made a time of amazing acceleration in molecular genetics tools—from the publication of whole genomes on the National Library of Medicine website to new technologies like RNA interference and fluorescence imaging to the rise of new private investors worldwide following web research reports, willing to take a risk. In the end, Elixir's ideas spawned several spinoff companies. Boston-based Ember Therapeutics focused on selective insulin sensitivity and the biology of brown fat, the good fat that burns calories. Agios Pharmaceuticals studied the growth factor pathway as a way of attacking cancer tumors. The company Rhythm Pharmaceuticals and other new companies focused on treating insulin resistance.

Through it all, Cynthia Kenyon and a few others promoted a vision despite the criticism. There is a progression in science, Leonard Hayflick once said. "First they say you're wrong. Then they say, well maybe. Then they say yeah it's true and everyone references your paper. Then it's so entrenched you're never referenced again." Kenyon saw that genes influence longevity and proclaimed it. "She deserves credit for sticking to the single gene idea," said Barzilai. "Otherwise people would have given up." It turned out the *FOXO* protective variant was present in various groups of human centenarians. The question was, what could be done with this knowledge?

An Aging Drug

By 2013, probably a dozen companies will be pursuing the longevity discoveries for possible drugs combating the diseases of aging. Worldwide, some 1,400 clinical trials are currently underway for new applications of rapamycin, 1,100 for metformin, 58 for resveratrol, and 11 for mitiglinide, a blood glucose–lowering drug that was the last of the Elixir interests. Elixir had a Huntington's compound in clinical trials in Europe, even though the company itself was defunct.

GlaxoSmithKline's Sirtris had lined up a large number of academic researchers who signed cooperative research agreements. Their new set of enzymes "at the very least make for complex and promising metabolic regulators," said Anne Brunet. Sirtris had three compounds in clinical testing. But with $720 million at stake, the pressure was rising. In the meantime, other company scientists may be researching sirtuins' effects as metabolic regulators. But few admitted to trying to slow aging. By the fall of 2012, reported Melinda Stubbee, GlaxoSmithKline director of media, global R&D, and pipeline news, "GSK isn't doing any research on longevity."

Insulin remains a main focus in many of the diseases of aging. As a modifier of insulin signaling, the widely prescribed diabetes drug metformin and other compounds were being tested for late-life health issues, including obesity, cancer prevention or treatment, and heart disease.

Easter Island's rapamycin, also known as sirolimus, is also being studied by up to a dozen pharmaceutical companies in clinical studies worldwide, both as an immune system suppressant and also for its potential in treating chronic disease. Many have studied its anticancer properties, including Yale University in partnership with Celgene Pharmaceuticals, which tested rapamycin for prevention of breast and other cancers. In Indiana, the Hoosier Biomedical group has studied it for treatment of renal cell carcinoma. The University of Chicago, with Genentech, has investigated rapamycin's antitumor activity in combination with another drug to treat late-stage cancers.

Resveratrol is under numerous clinical studies, including at the NIA for Alzheimer's. Worldwide studies on resveratrol for diabetes, obesity, cancer, and heart and brain health are ongoing. A University of South Australia study of twenty-eight obese men taking resveratrol showed that they had improved circulatory function. A Netherlands Maastricht University study of the compound showed improvement in the health of some eleven obese men by activating SIRT1 and the energy regulator called AMPK1. Another study by the same team showed similar improvement in a similar small number of obese subjects. Spain's National Research Council has studied resveratrol's possible cardioprotective role. The University of Arrhus in Denmark has studied its effect on insulin sensitivity and fat metabolism.

Several new companies are chasing other discoveries. Cardax Pharmaceuticals in Hawaii is focusing on the *FOXO* gene and inflammation. Proteostasis in California seeks dementia preventives in genes that prevented protein degradation. Cohbar is looking at the compound humanin's effect on mitochondrial peptides, the signals from the cell's energy factories. Several studies were testing CETP inhibitors to slow brain aging.

Meanwhile in the United States, the NIA budget has been cut. Only 5 percent went to the basic biology of aging. The agency invested in expensive clinical trials of compounds like resveratrol, crowding out the animal research that created the longevity gene revolution. "Those trials were what a drug company would do," said Richard Miller. "A lot of scientists were upset."

Into the breach poured private money. A Chinese energy company invested in Cohbar. A Russian plutocrat began yet another biotechnology fund. GlaxoSmithKline, mining the sirtuin pathway for drug treatments, remained a major player. But PayPal founder Peter Thiel, who funded the Methuselah Mouse, also helped fund Cynthia Kenyon's research. Dole chairman David Murdock created a billion-dollar North Carolina Research Campus to study fresh food and healthful aging. Arch Venture Partners contacted Judy Campisi about starting a company with the Mayo Clinic researchers of senescent cells. "I told them I have to be able to speak my mind," she said. The X Prize Foundation, funding the $10 million Archon Genomics X Prize to sequence the genomes of one hundred centenarians, chose Medco Health Solutions, a pharmacy benefit manager, to be its presenting sponsor. A few lab discoveries had ignited a science gold rush.

Two miles from the venerable Harvard campus and across the boulevard from MIT's modernist complex lies East Cambridge, once an unprepossessing, ramshackle, low-rise neighborhood. Today it looks more like a scene from the sci-fi movie *Gattaca*, with neon double helix signs and gleaming towers housing the brightest of the Bay State innovation economy—companies like Biogen, Vertex and Millennium Pharmaceuticals, not to mention Sirtris in its own nondescript office block. The same could be said of other biotech centers in California, Texas, North Carolina, and elsewhere. The final step is to ask, fifty years from now, what will people say of this moment in science history?

A Science Parable

The Copernican revolution in biology revealed the startling insight that we share most of the same genes and molecular pathways with every other living thing, from bacteria to kiwi to kangaroos. In life span science, we saw that we could learn a lot from aging in yeast, flies, mice, and worms. We learned that new tools could make it possible to tease out single genes that had a profound effect on length of life.

If you added up all the money invested in longevity genes over the last decade, it would total one-and-a-half to two billion dollars. Even the twentieth century's space and atomic races were by contrast mostly bureaucratic quests resembling military campaigns. Longevity was the individual imagination in small labs fighting with each other to penetrate the most intimate question of all, how long we will live. But what was it in actual science achievement?

"It was a profound transition point," said David Gems, "when we moved from description to cause and effect. No one else had Kenyon's originality and playful imagination. You know, these wacky ideas that no one else thought of, she'd go out and demonstrate them in a meticulous ironclad experiment."

Our understanding of the molecular genetics of longevity deepened as our lab technologies improved. "It was a tumultuous time," said David Sinclair. "It was high risk, high reward. People's lives were at stake, along with reputations, fortunes, and credit. In any groundbreaking science, there was going to be controversy. It's not easy. Aging's hard enough to control in a worm, let alone in a mouse or a human."

Out of the nexus of imagination, technology, ego, and venture capitalism came several compounds that looked promising in clinical trials. "I think we will have a drug to slow human aging," said Matt

Kaeberlein. "There is no reason to think we won't be successful, since evolution never really bothered with our life span after we reproduce," said Derek Lowe. "There is a whole lot of room to play. I just don't want to be the first to take the drug!"

If a science paradigm changes, Thomas Kuhn wrote, the world changes with it. New instruments seek new data by looking in the old places, and old instruments combine to look in new places. The effect of discovering the first longevity genes was like that of Midas's gold. A social explosion surrounded the science idea. The genes were seen to behave like investors, protecting their money and taking risks. Some scientists vied for media supremacy in a fight to advance their own interpretations.

On the one hand, biotech is what we have been doing on farms and greenhouses for thousands of years, seeking more productive or marketable animal and plant hybrids for society's needs and profit. Today we are doing this same genetics at unbelievable speed. Rewiring the life code is going to be the next industrial revolution, says Harvard Business School scholar Juan Enriquez. Companies may develop computer code to insert into our DNA. The company Ion Torrent is seeking to democratize the genome by bringing down the cost of our personal gene sequencing from $100,000 to about $7,500. Life Technologies in Carlsbad, California, is planning to offer a $1,000 personal genome by the end of 2013.

When it came to aging, some researchers and investors dreamt of a new alchemy to transmute longevity genes into money. New levels of funding from angel investors joined with new uses of science communication and an age-old public fascination. The science followed the first steps of the transformation outlined by Thomas Kuhn in *The Structure of Scientific Revolutions*. It had a status quo, a rebellion, and a protracted argument featuring disagreements over basic terminology. It is still an immature science, in its adolescent phase of rapid growth and stunning mistakes.

The insecurity and bravado that got them funded in the first place, that fueled the Internet binge too, drove some of the gerotechnology boom. "To get into the big journals, you had to be polished and well-spoken and David and Cynthia had that," said Rozalyn Anderson. "Reporters and editors are frankly a bit lazy. If you give them something polished, they're going to publish it the way you want them to."

The problem was scientists sometimes break their own rules, perhaps trying to satisfy our own longing, making their story a parable. The final question was, what wisdom did we gain?

My Family and Other Animals

It is New Year's on Marco Island, Florida. The wind tosses the warm waves on the Gulf. One hundred eighty miles away lies Cuba. The average age in my building is about seventy. Buffeted by the wind, I feel the proteins in each cell striving to retain their form, my insulin receptors boosting its signal, and the many mitochondria struggling. Suddenly, I see that that debilitating illness happens at any time. *Memento mori.* This is our future.

Many of us will live longer than our parents. If we are not too obese or stricken by disease, or lacking in medical care, we can hope to be healthier than in the past. Because many of us will live longer, we will need a little more money, or a little less expenditure. We can hope for continued technological progress to provide an ever-improving lifestyle, but it will be more important to save. We can hope to be happy, engaged, creative, and somewhat contributive. We can hope for a slight adjustment in Social Security to assure the next generations beyond ours of the same benefits we have enjoyed. Perhaps employers will enable us to seek out flexible work programs. We hope for an improving world.

The clouds are whipping over the gulf when my father asks if I want to walk to the boat park, about a quarter-mile down to the island's very end. My mother and wife are out shopping. It seems like a good idea.

Biogerontologists insisted we could look at animals that live without aging, keep growing and reproducing until they die, he said. Bowhead whales, alligators, Bradley's turtles, tortoises, and Patagonian toothfish, all lived much longer than humans. Other deep sea, coldwater fishes or the tubules at the hot or cold vents beneath the Gulf of Mexico, lived up to two hundred years. So there was nothing new about genetic control of healthful long life. We had just never tried it. To do so, we needed the worm.

My father carefully zips his red windbreaker. As we leave the condominium, we look out over the Ten Thousand mangrove islands, hearing the raucous call of an osprey over the wind. Here was probably where Ponce de Leon, commander for Columbus, inched along five hundred years ago, seeking gold and a putative fountain of youth. We make our way slowly, my father's pace set by his right foot, slowed by his stroke ten years ago, step, stop. Step stop.

C. elegans revealed some of the circuitry that determines our length of fitness. Worm in the apple, creature that feeds on decay, in our time it showed life span was plastic. The indestructible worm could live ten times its normal life span if you tweaked a single gene and an accompanying pathway. Generating a field of ten thousand researchers, it made the unlikeliest of heroes, a paradox of survival and destruction. It showed us a way potentially to vastly improve the human experience at the end of life.

At the boat park, my father plops down on a stone bench as the snook fishermen reel in their lines before the coming squall. I greet a boat captain who asks if I have watched the immortality documentary he recommended. The rising wind rattles the Coast Guard hut's shutters. The fishing and pleasure boats are sailing back into port.

The science of aging made a subversive show that brought something of infinity to the common world. It was subject to all the follies of any human endeavor, and gave us a science very unlike our popular impression of progress—chaotic, unbridled, greedy, idealistic, imaginative, emotional and, at times, even humorous. The problem was scientists sometimes committed mistakes in trying to satisfy our longing, making their story point to a future that, year by year, eludes us.

As the rain starts we head back as fast as my father can. Step, stop. Step, stop. It is incredibly slow. I remember when he taught me how to ride a bike, not knowing how himself, picking me up each time I fell over in our little driveway. He pauses to rest once again, gripping my hand. I see myself, holding onto my son or daughter, sometime in the future. The rain scuds and pours, the wind howls, and together we push ourselves, shaking and staggering a little, back into the lobby. He falls into the chair. "Ohhh-kaaay," he says. We live to fight another day.

It is the beginning of the new science of the molecular genetics of aging. The longevity genetics tools developed in the 2010s launched a young science into its adolescent age. "The next decade," promised Brian Kennedy, "is going to be amazing."

Epilogue

Aging is the coda of each of our stories. Our characters shape our genetic code, not the other way around. We hope for these additional years to permit us to fulfill ourselves, maybe attending to the delayed amends and risks we wished we had taken. Aging develops the soul as we shift from thinking about personal details to larger meanings, reflecting on and repeating actions as a sort of distillation process. As we glimpse the possibilities of a productive later life, we can glimpse what healthful longevity may mean for the individual, society, and the human experience.

The scientists of longevity genes took risks. They chose the right or wrong pathways, got lucky or unlucky, and ended up at the most amazing places imaginable. They sifted through the complexity and clutter of life, which was also the complexity and chaos of science history. They explored the unknown and sought to utilize its promise.

These scientists struggled to make sense of their discoveries. Their journeys went to the core of what it means to be human, the transmutation of nature into something akin to art. It is the beginning of a new science, not the end.

Where They Are Now

ROZALYN ANDERSON is an assistant professor in the Department of Geriatrics and Adult Development at the University of Wisconsin–Madison Institute on Aging and is studying the potential correlations between caloric restriction and age-associated disorders.

NIR BARZILAI is a professor in the Medicine and Genetics departments at the Albert Einstein College of Medicine, and is the Ingeberg and Ira Leon Rennert Chair of Aging Research, as well as founding director of the Institute for Aging Research at the Albert Einstein Medical Center. He was named to the "Forward 50," a list of people who have made a significant impact on the Jewish story.

HOLLY BROWN-BORG is a researcher at the University of North Dakota, where she is a professor of physiology, and is a past president of the American Aging Association and past chair of Biological Sciences of the Gerontological Society of America. Her current research is on the effects of increased and decreased growth hormone levels on the longevity of mice.

ANNE BRUNET is associate professor of genetics at Stanford, where she studies the genetic pathways that connect insulin to FOXO transcription factors to regulate life span from worms to humans. She is interested in the role of longevity genes in the maintenance of adult neural stem cells and intact cognitive function during aging. She also uses ultra-high throughput sequencing technologies to study epigenetic changes in chromatin modifiers in life span and metabolism.

ROCHELLE BUFFENSTEIN is a professor in the Department of Physiology at the University of Texas Health Science Center's Barshop Institute for Longevity and Aging Studies, where she is studying aging by observing naked mole rats.

JUDITH CAMPISI is a senior scientist who works on cancer and aging at Lawrence Berkeley and is a professor at the Buck Institute for Age Research. She is studying aging and cancer through research on mouse cells, human cells, and *C. elegans*.

ANDREW DILLIN is investigating insulin signaling, protein folding, and neurodegenerative disease in the aging process as professor in the University of California at Berkeley Molecular and Cell Biology Department.

MONICA DRISCOLL, a professor in the Department of Molecular Biology and Biochemistry at Rutgers University, is studying the composition and regulation of a new class of ion channels in *C. elegans*.

DAVID GEMS is a reader in the biology of aging and is deputy director of the Institute of Healthy Ageing at University College London, where he works on the genes and biochemical processes by which reduced insulin/IGF-1 signaling and dietary restriction increase life span. Other interests include sex differences in the biology of aging, evolutionary conservation of mechanisms of aging, and bioethical implications of aging research.

LEONARD GUARENTE is Novartis Professor of Aging at MIT, where he continues to explore the avenues by which genes similar to *SIR2* regulate life span.

F. BRADLEY JOHNSON is associate professor of pathology and laboratory medicine at the University of Pennsylvania and is currently studying the role that telomeres, chromatin, and cell senescence play in aging and cancer.

THOMAS E. JOHNSON researches longevity in mice and worms as a professor of behavioral genetics at the Institute for Behavioral Genetics on the campus of the University of Colorado at Boulder.

MATT KAEBERLEIN codirects the Nathan Shock Center of Excellence in the Basic Biology of Aging at the University of Washington. He is studying the mechanisms by which conserved longevity pathways modulate healthy aging.

BRIAN KENNEDY is now president and CEO of the Buck Institute for the Study of Aging in Marin County, California, where researchers investigate correlations between chronic disease and aging.

CYNTHIA KENYON is the Herbert Boyer Distinguished Professor and Director of the Hilblom Center for the Study of Aging at the University of California, San Francisco.

MICHAEL KLASS, after eighteen years at Abbot Labs in Chicago, served as vice president of research and development at Caris Life Sciences. He is currently president of MRK Consulting LLC, a private company in Tucson, Arizona, that focuses on development, regulatory compliance, and clinical validation of diagnostic products.

VALTER LONGO is the Albert L. and Madelyne G. Hanson Family Trust Professor of Gerontology and professor of biological sciences at the University of Southern California, Davis School of Gerontology. He studies the molecular pathways conserved from simple organisms to humans that can be modulated to protect against multiple stresses and delay or prevent Alzheimer's disease and other diseases of aging. His focus is on the pathways that regulate resistance to oxidative damage in yeast and mammalian neurons.

RICHARD MILLER is professor of pathology at the University of Michigan Medical School, director of the Nathan Shock Center in the Basic Biology of Aging, and associate director for research at the University of Michigan Geriatrics Center. He studies cell stress resistance and longevity in mice and various wild bird species.

RICK MORIMOTO researches proteins and healthy aging as the Bill and Gayle Cook Professor of Biology in Northwestern University's Department of Molecular Biosciences and the Rice Institute for Biomedical Research.

COLEEN MURPHY is an associate professor in the Department of Molecular Biology and Richard B. Fisher Preceptor in Integrative Genomics at Princeton. Her lab is focused on identifying the genes controlled by the global regulators of longevity and elucidating the biological and biochemical mechanisms used by these genes to affect life span.

S. JAY OLSHANSKY is a professor in the School of Public Health at the University of Illinois at Chicago, and a research associate at the Center on Aging at the University of Chicago and the London School of Hygiene and Tropical Medicine. He is investigating the public policy implications of increased human longevity.

LINDA PARTRIDGE is founding director of the Max Planck Institute on Aging in Cologne, Germany, and Weldon Professor of Biometry at the Department of Genetics, Evolution, and Environment at University College London. She focuses on the correlations between high rates of reproduction and rates of aging.

TOMAS PROLLA is professor of genetics and medical genetics at the University of Wisconsin–Madison School of Medicine and Public Health. He is focused on understanding the molecular basis of the aging process and age-related human diseases through the use of large-scale gene expression analysis. Having characterized the gene expression profiles of thousands of genes during aging in skeletal muscle and the brain, he is currently extending these studies to multiple tissues of mice, humans, and rhesus monkeys.

GARY RUVKUN is professor of genetics at Harvard Medical School and the Howard Hughes Medical Institute, where he studies RNAi pathways as well as mechanisms of aging and toxin surveillance.

DAVID SINCLAIR is a professor in the Harvard Medical School Department of Genetics, codirector of the Paul F. Glenn Laboratories for the Biological Mechanisms of Aging, and cofounder of Sirtris Pharmaceuticals. His research aims to identify conserved longevity control pathways and devise small molecules that activate them, with a view to preventing and treating diseases caused by aging. His lab discovered key components of the aging regulatory pathway in yeast and is now focused on finding genes that extend the healthy life span of mice.

P. ELINE SLAGBOOM, professor and department head of molecular epidemiology at the Leiden University Medical Center and cofounder of the Research on Aging Center, studies the human genetics of aging through the Leiden Longevity Study she co-initiated.

MARC TATAR, professor of ecology and evolutionary biology at Brown University, studies insulin signaling and longevity in fruit flies.

HEIDI TISSENBAUM is a professor in the Program in Molecular Medicine at the University of Massachusetts Medical School. She is investigating chemical pathways affecting longevity in *C. elegans* with the hope of applying any findings to the same pathways in humans.

Acknowledgments

I wish to thank the following individuals for their time and expertise. Without them, this book could not have been written. Any errors in the book, however, are purely my responsibility.

First, for crucial financial support, time, and thoughtful fellowship, I thank the Humanities Center of DePaul University, its director, Jonathan Gross, and associate director Anna Clissold. I thank the DePaul University Research Council and Competitive Research and Leave programs for providing the time and money to complete this work, chair Anne Bartlett, and science librarian Chris Parker for his geneous guidance.

For their expert assistance as researchers, editors, assistants, and collaborators, I thank Bryan Kett, Miranda Lukatch, and Dr. Elizabeth Stein, as well as several of my undergraduate interns who assisted in transcribing the hours of interviews.

For their careful reading as editors, I wish to thank David Groff, Howard B. Levi, Michele Morano, and Alexandra Reid. For their careful reading as scientists, I thank Andrzéj Bartke, Nir Barzilai, Tom Johnson, Matt Kaeberlein, Cynthia Kenyon, Michael Klass, Rick Morimoto, S. Jay Olshansky, Heidi Tissenbaum, and Bradley Willcox. Once again, I alone am responsible for the final product.

The following scientists, family members, investors, and reporters were interviewed, and I gratefully acknowledge their contributions: Rozalyn Anderson, Nuno Arantes-Oliveira, Joy Alcedo, Javier Apfeld, Roxanne Aune, Steven Austad, Father Nicanor Austriaco, Andrzéj Bartke, Cindy Bayley, Sydney Brenner, Holly Brown-Borg, Anne Brunet, Rochelle Buffenstein, Judith Campisi, Ed Cannon, Andrew Dillin, Michelle Dipp, Peter Distefano, Caleb Finch, David Friedman, David Gems, Leonard Guarente, Malene Hansen, Jeanne Harris, Leonard Hayflick, Stephen Helfand, Ira Herskowitz, Konrad Howitz, Shin Imai, Brad Johnson, Tom Johnson, Matt Kaeberlein, Erik Kapernik, Brian Kennedy,

Jane Kenyon, Michael Klass, Natasha Libina, Gordon Lithgow, Derek Lowe, Will Mair, George Martin, Richard Miller, Jon Moore, Jason Morris, Coleen Murphy, S. Jay Olshansky, Linda Partridge, Tomas Prolla, Jen Raymond, David Reznick, Tom Rando, Jasper Rine, Gary Ruvkun, Andrew Samuelson, Heidi Scrable, Phillip Sharp, David Sinclair, Eline Slagboom, Kristan Steffen, Marc Tatar, John Tower, Jan Vijg, George Vlasuk, Cindy Voisine, Nicholas Wade, David Waring, Alicyn Warren, Alan Watson, Rick Weindruch, Bradley Willcox, Lisa Wrischnik, and DePaul's Tom Donley. Others who gave of their time and expertise include Nick Bishop, Ehud Cohen, Rory Curtis, Julie Huber, Valter Longo, Justin Maloof, Rick Morimoto, and Jonathan Solomon. Other experts who talked to me include Anne D. Basting, Charles Harper, and Bill Heiden.

As scholars and colleagues who introduced me to the new field of age studies, I thank Erin Lamb, Leni Marshall, Teresa Mangum, Michelle Massie, and Cynthia Port.

My editor, Christie Henry, copy editor Mary Frances Gehl, and associate editor Amy Krynak, along with promotions manager Melinda Kennedy, promotions director Levi Stahl, and designer Matt Avery, provided critical assistance and wisdom. For guidance, patience, and support with this project through its many drafts from the very beginning, I thank my literary agent, Ellen Levine.

Finally, for all of their camaraderie and good wisdom, I thank my parents, Gus and Bertha, my children, Marja and Cam, and my wife, Maja, model and muse.

Longevity Gene Timeline

November 1, 1934	Clive McCay discovers that calorie restriction extends rats' life span
July 1, 1979	Tom Kirkwood proposes the disposable soma aging theory
February 5, 1988	Tom Johnson discovers that the *age-1* gene extends life span up to 50 percent in nematodes
September 15, 1993	Cynthia Kenyon finds that mutations in *daf-2* double nematode life span
February 10, 1995	Brian Kennedy and colleague find that *SIR4-42* extends yeast life span by 30 percent
1996–1997	Jim Thomas and Gary Ruvkun clone and analyze *age-1*
October 15, 1997	Gary Ruvkun discovers that *daf-2* encodes a receptor similar to human insulin and insulin-like growth factor receptor
December, 1997	Kenyon and Ruvkun clone and analyze *daf-16*, the long-life transcription factor regulated by *daf-2*
December 1999	Matt Kaeberlein and Leonard Guarente discover that *SIR2* extends yeast life span by 40 percent
March 2002	Coleen Murphy and Jim Thomas discover that *daf-16* controls one hundred antistress, antimicrobial, proimmune, and antioxidant genes in the nematode
September 2002	Martin Holzenberger discover that female mice lacking insulin receptors live 30 percent longer than normal
January 24, 2003	Matthias Bluher, Barbara Kahn, and Ronald Kahn show that male and female mice lacking the insulin receptor in fat tissue live 18 percent longer than normal
August 23, 2003	David Sinclair finds that resveratrol activates SIRT1 and extends life span in yeast

November 18, 2005	Matt Kaeberlein, Brian Kennedy, and others discover that the mTOR gene extends yeast life span
November 2007	Jill Milne and Sirtris researchers discover two putative small molecule activators of sirtuins that are more powerful than resveratrol
June 2008	GlaxoSmithKline buys Sirtris for $720 million
September 2008	David Harrison and David Sharp discover that rapamycin extends adult mouse healthful life span
October 2008	Brad Willcox, Craig Willcox, Nir Barzilai, and Yousin Suh find that centenarian populations share FOXO3a gene mutation
December 2009 and January 2010	Journal articles by Amgen and Pfizer scientists challenge the role of resveratrol or its mimetics in triggering sirtuins
September 2011	David Gems, Linda Partridge, Matt Kaeberlein, and colleagues suggest that sirtuins do not extend life span significantly in flies or worms
November 2011	Anne Brunet shows that an epigenetic longevity gene change can be inherited for up to four generations in a worm
November 2011	Jan van Deuersen and colleagues at the Mayo Clinic demonstrate that aging cells' removal extends the life span of obese mice

List of Longevity Genes

age-1: Regulated by *daf-2*, this worm gene is also known as *daf-23*. It encodes a PI3 kinase, an enzyme involved in cellular functions such as cell growth, proliferation, differentiation, survival, and intracellular trafficking, also involved in cancer prevention and life span.

AMP-kinase: Energy sensor and regulator controlled by sirtuins and FOXO3a, making an enzyme that plays a role in cellular energy balance, extending life.

daf-2: A gene for the insulin and insulin-like growth factor receptor in the worm *C. elegans*, controls aging, among many other life processes. Cynthia Kenyon demonstrated that mutations in this gene have the potential to double the life span of *C. elegans*.

daf-16: In the same pathway as *daf-2*, *daf-16* in *C. elegans* makes a transcription factor that controls numerous processes of aging as well as the response to environmental stimuli. Numerous species, including humans, have genes similar to *daf-16*.

FOXO3a: The human version of *daf-16*, also found in many different species. In humans, the gene encodes a protein that is believed to be affiliated with apoptosis, or, programmed cell death, and healthy longevity.

HSF-1: When temperatures are increased in an organism, this gene encodes a transcription factor to protect protein folding.

IGF1: The protein encoded by this gene is one of the key drivers of human growth. Down regulating increases life span in mice and other animals.

JNK1: A gene making a protein activated by environmentally stresses, such as UV radiation. Inhibition of this protein increases sensitivity to insulin, and in obese mice and other organisms, the activity of this protein is increased with a longer life span.

Klotho: Mice deficient in this gene exhibit a syndrome resembling premature aging.

mitochondrial peptides: Nir Barzilai discovered that when infused in diabetic rats, mitochondrial peptides increased insulin sensitivity. These signalers from mitochondria to nucleus are regulated by the gene JNK1.

mTOR: Mammalian target of rapamycin, a gene in a chemical signaling pathway that, when mutated, is believed to prolong life through caloric restriction.

Nrf2: In mice and African naked mole rats, this gene is involved in cancer resistance. Also present in humans and other animals, it activates many cytoprotective proteins, defending cells against harmful agents.

P66Shc: In mice, if this gene was mutated, stress resistance to environmental factors, such as ultraviolet light, increased. This in turn prolonged the life of the mice.

pha-4: When *pha-4* was removed from worms, any life extension from caloric restriction was negated.

PNC1: A part of the NAD signaling pathway in yeast, necessary for the life extension induced by caloric restriction.

Pit1: This gene encodes a transcription factor that controls the production of growth hormone in mice.

Sir2: The first type of sirtuin found, making a class of protein that regulates aging in yeast, the replication of DNA, and apoptosis. It was believed that SIR2 played a role with extended life and caloric restriction.

SIRT1-7: The human version of *Sir* genes. Humans have seven versions, of which SIRT1, contributing to the process of aging and extending life, is the most studied. The other six versions of human sirtuin genes are under exploration. *SIRT2* plays a role in cell division, genome integrity, and tumor prevention. *SIRT3-5* may act in the mitochondria and assist in metabolic processes. The *SIRT6* protein has a role in postnatal development, inflammation reduction, and cancer protection. The overexpression of *SIRT6* can lengthen the life of male mice by some 15 percent. *SIRT7* appears to work mostly in heart cells, regulating cell death and stress response and rDNA transcription, and is involved in tumor formation, heart health, and chromatin maintenance.

skn-1: In *C. elegans*, *skn-1* encodes a transcription factor that influences stress resistance and, in parallel with *daf-16*, regulates life span.

Notes

PREFACE

xi *yearned to do something great* Cynthia Kenyon, phone interview, March 13, 2008. Unless otherwise noted, all quoted conversations are based on personal interviews conducted by the author or public documents.

xi *pulled over a microscope to take a closer look* Cynthia Kenyon, phone interview, May 23, 2001.

xii *one in three people in the developed world* "World Population Ageing: 1950–2050," Population Division, *DESA*, United Nations, accessed July 7, 2012, http://www.un.org/esa/population/publications/worldageing19502050/pdf/80chapterii.pdf.

xiii *the intensity of this "silver tsunami"* *Global Population Ageing: Peril or Promise?* (New York: World Economic Forum Global Council on an Aging Society, 2012), 4.

xiv *in ancient Rome, people lived twenty-nine years* Roger Bagnall and Bruce Frier, *The Demography of Roman Egypt* (Cambridge: Cambridge University Press, 1994), 37.

CHAPTER 1

3 *pleiotropy means one gene has many effects* George Williams, "Pleiotropy, Natural Selection, and the Evolution of Senescence." *Evolution* 11 (1957): 398–411.

4 *vasectomies would increase male longevity* Pope Brock, *Charlatan: America's Most Dangerous FlimFlam Man, the Man Who Pursued Him, and the Age of Flim-Flam* (New York: Crown, 2009), 207.

4 *Brinkley was so successful* Ibid., 90.

4 *banishes the 'fountain of youth' to the limbo* Williams, "Pleiotropy, Natural Selection, and the Evolution of Senescence," 402.

5 *McCay published his longevity findings* Clive McCay and Mary F. Crowell, "Prolonging the Life Span," *Scientific Monthly*, November 1934, 405–14.

5 *in a best-selling book called* Maximum Life Span Roy Walford, *Maximum Life Span* (New York: Norton, 1983).

6 *the American NIA pushed academic science* Robert H. Birnstock, "The Search

for Prolongevity: A Contentious Pursuit," in *The Fountain of Youth: Cultural, Scientific, and Ethical Perspectives on a Biomedical Goal*, ed. Stephen Post and Robert Birnstock (Oxford: Oxford University Press, 2004), 14.

6 *if we found one thing, a trick say* Ron Hart, quoted in Carol Kahn, *Beyond the Helix: DNA and the Quest for Longevity* (New York: Times Books, 1985), 5.

6 *as he sat in the steaming water, Kirkwood realized* Tom Kirkwood, *Time of Our Lives: The Science of Human Aging* (New York: Oxford University Press, 1999) 63–65.

6 *the proposition that* longevity *is somehow determined* Leonard Hayflick, "Origins of Longevity," in *Modern Biological Theories of Aging*, ed. Huber Warner (New York: Raven Press, 1987), 21 (original emphasis).

7 *they parasitize almost everything we eat* Andrew Brown, *In the Beginning Was the Worm: Finding the Secrets of Life in a Tiny Hermaphrodite* (New York: Columbia University Press, 2003), 30–32.

7 *like seeing God* Joy Alcedo, interview with author, July 18, 2002.

8 *my head was reeling* John White, phone interview with author, March 29, 2001.

8 *gang of cryptographers* Letter from Joshua Lederberg to Sydney Brenner, June 14, 1968,original copy, June 28, 1958, collection of the author.

8 *"fixed entities" like genes* Carl Woese, "Bacterial Evolution," *Microbiological Reviews* 51 (June 1987): 223.

8 *we were an evangelical sect* Sydney Brenner, interview with author, May 21, 2001.

8 *as an unending argument* Freeman Dyson, "The Dramatic Picture of Richard Feynmann," *New York Review of Books*, July 14, 2011.

8 *flow of information* Sydney Brenner, interview with author, May 21, 2001.

9 *aging just seemed to me to be an illogical process* Michael Klass, phone interview with author, December 1, 2003.

10 *discovery landed them with a paper* Michael R. Klass and David Hirsh, "Non-Aging Developmental Variant of *C. elegans*," *Nature* 260 (1997): 523–25.

10 *specific life-span genes are extremely rare* Michael Klass, "A Method for Isolation of Longevity Mutants in the nematode *C. elegans*," *Mechanism of Aging and Development* 22 (1983): 286.

10 *he was thinking of leaving academia* Michael Klass, personal communication to author, April 29, 2012.

11 *I learned early on how to wait* Tom Johnson, phone interview with author, November 19, 2003.

11 *intractable scientific problem* Tom Johnson, phone interview with author, July 20, 2011.

12 *it was a pursuit of truth* David Friedman, phone interview with author, May 3, 2012.

12 *people looked at the result very skeptically* Ibid.

12 *to their mutual disbelief* David Friedman and Tom Johnson, "A Mutation

in the *age-1* Gene in *Caenorhabditis elegans* Lengthens Life and Reduces Hermaphroditic Fertility." *Genetics* 118 (1988): 75–86.

12 *Tom Johnson was interviewed* Tom Johnson, personal communication to author, May 15, 2012.

12 *Johnson traced the mechanisms* Tom Johnson, "Aging Can Be Genetically Dissected into Component Processes Using Long-Lived Lines of *Caenorhabditis elegans*." *Proceedings of the National Academy of Sciences U S A* 84 (1987): 3777–81.

13 *every time I appeared on TV* Tom Johnson, interview with author, August 2, 2010.

13 *DNA is Midas's gold* Horace Freeland Judson, *The Eighth Day of Creation: Makers of the Revolution in Biology* (New York: Cold Spring Harbor, 1996), 51.

CHAPTER 2

14 *something that everyone thought occurred passively* Cynthia Kenyon, interview with author, December 29–30, 2008.

15 *worried that she had experienced the best* Ibid.

15 *competing against the William Buckley family* Jane Jarvis Kenyon, phone interview with author, June 2, 2002.

16 *the family escaped to her father's parents' farm* Susan Fitch Kelsey, phone interview with author, June 3, 2002.

16 *catch a wild bird* Jane Jarvis Kenyon, phone interview with author, January 22, 2010.

16 *she became a "nature girl"* Alicyn Warren, phone interview with author, May 6, 2009.

17 *her father would drive along* Ibid.

17 *he drove her hard* Ibid.

17 *Cynthia Kenyon worked two jobs* Cynthia Kenyon, phone interview with author, April 28, 2004; Cynthia Kenyon, interview with author, September 28, 2009.

17 *reading chapters like "The Mendelian View of the World"* James Watson, *Molecular Biology of the Gene*, 1st ed. (New York: Cold Spring Harbor, 1977), 3.

17 *I thought it was the key* Cynthia Kenyon, phone interview with author, July 8, 2011.

18 *writing up her experiments in prominent journals* Cynthia Kenyon, "If Birds Can Fly, Why Can't We?" *Cell* 76 (July 1994): 175–80; Kenyon, "A Perfect Vulva Every Time," *Cell* 82, no.2 (July 1995): 297–307.

18 *she took them to the San Francisco opera* Jeanne Harris, phone interview with author, May 12, 2004.

19 *sketched out my career for me on a napkin* David Waring, phone interview with author, June 2, 2004.

19 *if your experiment was not working* Lisa Wrishnik, phone interview with author, June 1, 2004.

19 *the genome is built in such a way* Cynthia Kenyon, phone interview with author, January 22, 2001.

19 *the great paradigm shift in biology* Cynthia Kenyon, "The First Long-Lived Mutants: Discovery of the Insulin/IGF-1 Pathway for Aging." *Philosophical Transactions of the Royal Society of London B: Biological Sciences* 366 (2011): 9–10.

20 *he planned to treat them at 20° Celsius* Cynthia Kenyon, phone interview with author, August 1, 2010.

20 *the discovery opened a "big door"* Ira Herskowitz, interview with author, July 18, 2002.

21 *triggering expression of a regulated life span mechanism* Cynthia Kenyon, Jean Chang, Erin Gensch, Adam Rudner, and Ramon Tabtiang, "A *C. elegans* Mutant that Lives Twice as Long as Wild Type." *Nature* 366, no. 6454 (1993): 461.

21 *everything we know about the evolution of aging* Linda Partridge and Paul Harvey, "Methuselah among Nematodes," *Nature* 366, no. 6454 (1993): 429

21 *on the edge of something* Cynthia Kenyon, phone interview with author, May 23, 2001.

22 *kind of a dumb result* Gary Ruvkun, interview with author, June 12, 2008.

22 *could affect the life span of a mammal* George Martin, phone interview with author, January 29, 2001.

22 *people go all their scientific lives* Andrew Dillin, interview with author, December 10, 2010.

22 *a mixture of the workmanlike and the beautiful* David Gems, phone interview with author, June 9, 2010; Cynthia Kenyon, "Ponce d'elegans: Genetic Quest for the Fountain of Youth. *Cell* 84, no.4 (1996): 501–4.

22 *it was interesting* Gary Ruvkun, interview with author, June 12, 2008.

CHAPTER 3

23 *I lost a year and almost died* Brian Kennedy, phone interview with author, June 8, 2008.

23 *after three surgeries* Brian Kennedy, phone interview with author, March 11, 2011.

24 *Austriaco mentioned apoptosis* Nic Austriaco, phone interview with author, October 27, 2008.

24 *like something that just happened* Leonard Guarente, *Ageless Quest: One Scientist's Search for Genes That Prolong Youth* (New York: Cold Spring Harbor, 2002), 5.

24 *aging is in no sense any basic feature* "Michael Rose: Beating Death," *Discover*, May 1, 2001, accessed July 7, 2012, http://discovermagazine.com/2001/may/breakdialogue/.

25 *it seemed like aging in yeast* Brian Kennedy, phone interview with author, March 11, 2011.

25 *intoxicated by this thought* Guarente, *Ageless Quest*, 6.

25 *a good mentor for students with a risky dream* Leonard Guarente, phone interview with author, January 18, 2001.

26 *Guarente's postdoctoral work there* Leonard Guarente, interview with author, March 19, 2002.

26 *they discussed the problem* Brian Kennedy, phone interview with author, June 8, 2008.

27 *it took them four years rather than one* B. Kennedy, N. Austriaco, J. Zhang, and L. Guarente, "Mutation in the Silencing Gene SIR4 Can Delay Aging in *Saccharomyces cerevisiae*. *Cell* 80 (1995): 485–96.

27 *why go off into this field* Philip Sharp, phone interview with author, June 26, 2012.

28 *Guarente traveled to Australia* David Stipp, *The Youth Pill: Scientists at the Brink of an Anti-Aging Revolution* (New York: Current, 2010), 177–79.

28 *just needed one gene intervention* David Sinclair, phone interview with author, September 10, 2003.

28 *ideas and debates flying through the air* Brad Johnson, quoted in Jennifer Couzin, "Aging Research's Family Feud," *Science*, February 27, 2004, 33.

28 *he annoyed people* Ibid.

28 *Sinclair burst into Guarente's office* David Sinclair, phone interview with author, July 15, 2004.

29 *a precise molecular cause of aging* David Sinclair and Leonard Guarente, "Extrachromosomal rDNA Circles—A Cause of Yeast Aging." *Cell* 91 (1997): 1033.

29 *struggled to figure out the mechanisms* Leonard Guarente, interview with author, August 28, 2003.

29 *sequences very similar to SIR2* Guarente, *Ageless Quest*, 47.

29 *spanning a billion years of evolution* Leonard Guarente, phone interview with author, June 3, 2002.

29 *a fire drill threatened to stop a key procedure* Shin Imai, interview with author, February 21, 2006.

29 *medieval time travelers* Guarente, *Ageless Quest*, 47.

30 *something terribly wrong* Shin Imai, interview with author, September 12, 2009.

30 *Imai wrote it up in his diary* Ibid.

30 *a biochemical mechanism to set the pace of aging* S. Imai, C. Armstrong, and L. Guarente, "Transcriptional Silencing and Longevity Protein Sir2 Is an NAD-Dependent Histone Deacetylase," *Nature* 403, no. 6771 (2000): 795–800.

31 *tugging with gusto* Nicholas Wade, "Scientist at Work: Searching for Genes to Slow the Hands of Time," *New York Times*. September 26, 2000.

31 *if all the genes in a genome were switched on* Nicholas Wade, *LifeScript: How the Human Genome Discoveries Will Transform Medicine and Enhance Your Health* (New York: Simon and Schuster, 2001), 161.

32 *evolution overengineers for survival* Judith Campisi. phone interview with author, June 22, 2010.

32 *they were all upside down* Cynthia Kenyon, "Ponce d'elegans: Genetic Quest for the Fountain of Youth." *Cell* 84, no. 4 (1996): 501–4.

32 *the new molecular study of aging* Leonard Guarente and Cynthia Kenyon, "Genetic Pathways that Regulate Ageing in Model Organisms. *Nature* 408, no. 6809 (2000): 255.

32 *an intriguing jigsaw puzzle* Editors' letter, *Nature* 408, no. 6809 (2000): 223.

CHAPTER 4

33 *knocked my socks off* "The Rightful Place with Dame Linda Partridge," Partridge interview by Roger Bingham, *The Science Network Royal Society*, May 10, 2010, accessed July 22, 2012, http://thesciencenetwork.org/programs/the-rightful-place/the-rightful-place-with-dame-linda-partridge.

33 *former Gulf of Mexico oil rig supervisor* Cindy Bayley, interview with author, August 5, 2002.

33 *a chemistry student and track star* Tomas Prolla, phone interview with author, September 9, 2004.

33 *how many life-span genes will be found to exist* Cynthia Kenyon, "Ponce d'elegans: Genetic Quest for the Fountain of Youth." *Cell* 84, no. 4 (1996): 504.

33 *something manic about her optimism* Jeanne Harris, phone interview with author, May 12, 2004.

34 *now I'm in aging research* Gary Ruvkun, quoted in David Stipp, *The Youth Pill: Scientists at the Brink of an Anti-Aging Revolution* (New York: Current Books, 2010), 85.

34 *"extended exploration" of an anomaly* Thomas S. Kuhn, *The Structure of Scientific Revolutions* (Chicago: University of Chicago Press, 1996), 54.

34 *as a child, Ruvkun loved reading* Gary Ruvkun, interview with author, March 19, 2002.

34 *he hitchhiked around Central and South America* Ingfei Chen, "The Drifter: A Sinuous Pathway Has Led Gary Ruvkun to Probe How *C. elegans* Grows Old." *Science of Aging Knowledge Environment* 42 (2002): nf11.

35 *he really cared about us* Heidi Tissenbaum, phone interview with author, October 18, 2008.

35 *Gary taught us to be fearless* Jason Morris, phone interview with author, August 6, 2010.

35 *try not to push it so far* Gary Ruvkun, interview with author, June 24, 2008.

35 *the hottest field in biology* Gary Ruvkun, quoted in "Diet, Aging and Metabolism," *The Science Network* Sept. 10, 2009, accessed July 6, 2012, http://thesciencenetwork.org/programs/diet-aging-and-metabolism/gary-ruvkun.

35 *all these people were so far ahead of me* Heidi Tissenbaum, interview with author, October 17, 2002.

36 *to find the Grim Reaper gene* Ibid.

36 *she was pregnant* Heidi Tissenbaum, phone interview with author, August 1, 2010.

36 *as they are ratcheting by* Gary Ruvkun, quoted in Nicholas Wade, "A Worm and a Computer Help to Illuminate Diabetes," *New York Times*, December 30, 1997.

37 *the key to slowing the aging process* K. Kimura, H. Tissenbaum, Y. Liu, and G. Ruvkun, "*daf-2*, an Insulin Receptor-Like Gene that Regulates Longevity and Diapause in *C. elegans*." *Science*, August 15, 1997, 942–46.

37 *excitement surrounding the idea* Sean Curran, "Conserved Mechanisms of Life Span Regulation and Extension in *Caenorhabditis elegans*," in *Life Span Extension: Single Cell Organisms to Man*, ed. Christian Sell, Antonello Lorenzini, and Holly M. Brown-Borg, (New York: Springer Press, 2009), 36.

37 *it was a race* Gary Ruvkun, interview with author, June 24, 2008.

38 *genes encoding nutrient sensors regulate ageing* Cynthia Kenyon, "The First Long-Lived Mutants: Discovery of the Insulin/IGF-1 Pathway for Aging. *Philosophical Transactions of the Royal Society of London B: Biological Sciences* 366 (2011): 9–10.

38 *sent their report to the journal* Science Kui Lin, Jennie B. Dorman, Aylin Rodan, and Cynthia Kenyon. "*daf-16*: An HNF-3/Forkhead Family Member that Can Function to Double the Life-Span of *Caenorhabditis elegans*." *Science*, November 14, 2997, 1319–22.

38 *Ruvkun rushed his to* Nature S. Ogg, S. Paradis, S. Gottlieb, G. I. Patterson, L. Lee, H. A. Tissenbaum, and G. Ruvkun, "The DAF-16 Fork Head-Related Transcription Factor Transduces *C. elegans* Insulin-Like Metabolic and Longevity Signals." *Nature* 389 (1997): 994–99.

38 *one of the most cited papers* Tom Johnson, "A Sub-Field History: *C. elegans* as a System for the Analysis of the Genetics of Aging," *Science of Aging Knowledge Environment* 34 (2002): 4

38 *we're a dysfunctional family* Gary Ruvkun, interview with author, March 19, 2002.

39 *an explosion of analysis* Huber Warner, "Developing a Research Agenda in Biogerontology: Basic Mechanisms," *Science of Aging Knowledge Environment* 2005, no. 44 (2005): 33.

40 *telling you what you can't think* Gary Ruvkun, quoted in Nicholas Wade, "Tests Begin on Drugs that May Slow Aging," *New York Times* Aug. 17, 2009, accessed July 3, 2012, http://www.nytimes.com/2009/08/18/science/18aging.html?pagewanted=all.

40 *try the reproduction experiment* Cynthia Kenyon, interview with author, December 30, 2008.

41 *a plate with a worm that should have been dead* Javier Apfeld, interview with author, August 8, 2003.

41 *if it was taken from the intestines* Javier Apfield and Cynthia Kenyon, "Cell Non-Autonomy of *C. elegans* daf-2 Function in the Regulation of Diapause and Lifespan," *Cell* 95, no. 2 (1998): 199–210.

41 *they extended healthful life* Honor Hsin and Cynthia Kenyon, "Signals from

the Reproductive System Regulate the Lifespan of *C. elegans*." *Nature* 399, no. 6734 (1999): 362–66.

41 *see something we weren't meant to see* Cynthia Kenyon, phone interview with author, April 30, 2002.

42 *Lenny really begged me to ask* Heidi Tissenbaum, phone interview with author, May 27, 2011.

42 *Hsin was fifteen* Cynthia Kenyon, "My Adventures with the Fountain of Youth," *Harvey Lecture Series* 100 (2006), p. 55.

43 *most important, most wondrous map* Bill Clinton, quoted in ABCNews.com, "Genome Project Complete," June 26, 2000, accessed July 3, 2012, http://abcnews.go.com/Technology/story?id=99380&page=1.

CHAPTER 5

44 *the year opened a "biotech century"* Jeremy Rifkin, *The Biotech Century: Harnessing the Gene and Remaking the World* (New York: Tarcher/Putnam, 1998), iv.

44 *to harness a natural resource* Cindy Bayley, interview with author, January 28–29, 2002; August 12, 2003.

45 *share longevity discoveries from many fields* Science of Aging Knowledge Environment archive site, accessed January 5, 2012, http://sageke.sciencemag.org/.

45 *described the gold rush* R. John Davenport and Jennifer Toy, "Gero-Tech Sprouts, But Will It Bloom?" *Science of Aging Knowledge Environment* 28, (2002): ns6, accessed July 3, 2012, http://sageke.sciencemag.org/cgi/content/full/2002/28/ns6.

45 *grinding up rats and looking at their arteries* Judith Campisi, phone interview with author, June 14, 2011.

46 *influence of money on research* Drummond Rennie, quoted in Daniel Greenberg, *Campus Capitalism: The Perils, Rewards and Delusions of Campus Capitalism*. (Chicago: University of Chicago Press, 2007), 250.

47 *animals resisted oxidative stress* Joelle Dupont and Martin Holzenberger, "IGF Type 1 Receptor: A Cell Cycle Progression Factor That Regulates Aging" *Cell Cycle* (2003): 269–71.

47 *thus placing the mouse* Greg Critser, *Eternity Soup: Inside the Quest to End Aging*, (New York: Harmony Books, 2010), 75.

47 *Ames mouse, along with Snell dwarfs and GHR knockouts* Andrzéj Bartke, "Can Growth Hormone (GH) Accelerate Aging? Evidence from Transgenic Mice," *Neuroendocrinology* 78 (2003): 210–16.

48 *single biggest longevity breakthrough* Richard Miller, phone interview with author, December 13, 2011.

48 *triggers were growth hormone receptors* Andrzéj Bartke, phone interview with author, July 9, 2010.

48 *researchers, funded in part by Purina* Sutter et al., "A Single IGF1 Allele Is a Major Determinant of Size in Small Dogs," *Science*, April 6, 2007, 112–15.

49 *I try not to take a dogmatic position* Andrzéj Bartke interview by Roger Bingham, *The Science Network*, September 11, 2009, accessed July 2, 2012, http://thesciencenetwork.org/programs/diet-aging-and-metabolism/andy-bartke.

49 *if you spent an iota on the biology* Richard Miller, interview with author, June 27, 2008.

49 *insulin is a key player* Marc Tatar, phone interview with author, September 20, 2005.

50 *a glimpse of the control panel* Richard Miller, "Extending Life: Scientific Prospects and Political Obstacles." *Milbank Quarterly* 80, no.155 (2002):80.

50 *discoveries "narrowed the field"* Huber Warner, "Developing a Research Agenda in Biogerontology: Basic Mechanisms," *Science of Aging Knowledge Environment* 44 (2005): pe33.

50 *If it's true in flies and mice* Cynthia Kenyon, interview with author, July 17, 2002.

50 *We counted twenty-two labs* Coleen Murphy, interview with author, July 17, 2002.

50 *a new scientific tool, RNA interference* Andrew Pollack, "RNA Trades Bit Part for Starring Role in the Cell," *New York Times*, January 21, 2003.

50 *going to find a fountain of youth* Williams, quoted in Nicholas Wade, *Lifescript: How the Human Genome Discoveries Will Transform Medicine and Enhance Your Health* (New York: Simon and Schuster, 2001), 150

51 *everywhere, signs of aging* Leonard Guarente, interview with author, August 13, 2003.

52 *nature discovered the backbone* Miller, "Extending Life," 84.

54 *where should we set it* Leon Kass, "L'chaim and Its Limits: Why Not Immortality?" *First Things*, May 2001, 37.

54 *evolution of life span among wild opossums* Steven Austad, phone interview with author, February 9, 2005.

55 *if current trends continue* S. Jay Olshansky, "Duration of Life: Is there a Biological Warranty Period?," PCBE Transcripts, December 12, 2002, accessed June 20, 2012, http://bioethics.georgetown.edu/pcbe/transcripts/dec02/session2.html.

55 *we have been wildly successful* Ibid.

55 *it seemed like Kass did not understand* S. Jay Olshansky, phone interview with author, January 9, 2009.

55 *some of these long-lived mutants* Olshansky, "Duration of Life."

56 *regardless of whether you and I want it to happen* Ibid.

CHAPTER 6

57 *water dripped down from the wall air conditioner units* David Sinclair, phone interview with author, September 10, 2003.

57 *several other experiments going* David Sinclair, phone interview with author, November 23, 2011.

58 *David, don't say things like that* Rozalyn Anderson, phone interview with author, December 16, 2009.

58 *not a drug discovery motivation* Konrad Howitz, phone interview with author, November 13, 2011.

58 studying all the plant polyphenols Konrad Howitz, phone interview with author, September 25, 2003.

59 *it was a new suburb* David Sinclair, phone interview with author, June 8, 2005.

59 *I thought of dying, of having my parents die* Ibid.

59 *you did not have to understand everything* Ibid.

60 *the exact moment when his life changed* David Sinclair, phone interview with author, September 10, 2003.

60 *I was lying, David* Gary Ruvkun, "Energy, Metabolism and Life Span," presentation at Harvard Healthy Life Span Conference, Boston, MA, June 23, 2008.

60 *forget about life span* Cindy Bayley, interview with author, August 17, 2003.

61 *the guy from Biomol kept calling* Rozalyn Anderson, phone interview, December 16, 2009.

61 *got to find out what A, B, and C [are]* David Sinclair, phone interview with author, September 10, 2003.

61 *bind to and make it work faster* Tom Bearden, "Extended Interview: Dr. David Sinclair," *PBS Newshour*, February 1, 2005, accessed July 9, 2012, http://www.pbs.org/newshour/bb/science/jan-june05/aging-sinclair_ext.html?print.

61 *something very big* David Sinclair, phone interview with author, September 10, 2003.

61 *organic substance derived from knotweed* Linus Pauling Institute, "Micronutrient Information Center: Resveratrol," updated June 2008, accessed July 7, 2012, http://lpi.oregonstate.edu/infocenter/phytochemicals/resveratrol/.

61 *few other companies* Peter Distefano, interview with author, August 18, 2003.

62 *been working on this all summer* Marc Tatar, phone interview with author, November 19, 2003; July 6, 2005.

63 *my flies are living longer* Ibid.

63 *put the news on page two* Rick Weiss, "Scientists Find Way to Stimulate Anti-Aging Enzyme," *Washington Post*, August 24, 2003, 2.

64 *activator of a known life-extending enzyme* K. T. Howitz, K. J. Bitterman, H. Y. Cohen, D. W. Lamming, S. Lavu, J. G. Wood, R. E. Zipkin, "Small Molecule Sirtuin Activators that Extend *S. cerevisiae* Lifespan," *Nature* 425 (2003): 191–96.

64 *three hundred other molecules* Julie L. Huber, Michael W. McBurney, Peter S. Distefano, Thomas McDonagh, "SIRT1-Independent Mechanisms of the Putative Sirtuin Enzyme Activators SRT1720 and SRT2183," *Future Medicinal Chemistry* 2, no. 12 (2010); 1751–59.

64 *been waiting for this all my life* Sinclair, quoted in Nicholas Wade, "Life Extending Molecule is Found in Certain Red Wines," *New York Times*, August 24, 2003.

64 *a yeast cell growing on a grape* Konrad Howitz and David Sinclair, "Xenohormesis: Sensing the Chemical Cues of Other Species." *Cell* 133 (2008): 386–91.

64 *overexpressing the gene* Stephen Helfand and Blanka Rogina, "Sir2 Mediates Longevity in the Fly through a Pathway Related to Calorie Restriction," *Proceedings of the Natl. Academy of Sciences U S A* 101 (2004): 15998–6003.

64 *a particular Biomol lab assay or test* Heidi Ledford, "Much Ado about Ageing," *Nature* 464 (2010) 480–81.

64 *media drive the hype* Judith Campisi, phone interview, June 22, 2010.

65 *close to a miraculous molecule* David Sinclair, quoted in Jennifer Couzin, "Aging Research's Family Feud," *Science*, February 27, 2004, 1297.

65 *Howitz and Sinclair sought financial backing* Konrad Howitz, phone interview with author, November 13, 2011.

65 *doctors with little interest in business* Toby Stuart and David Kiron, "Sirtris Pharmaceuticals: Living Healthier, Longer," Harvard Business School Case Study N9–808–112 (March 18, 2008), 4.

66 *sign a nondisclosure agreement* David Stipp, "Drink Wine and Live Longer" *Fortune*, February 12, 2007, 4.

66 *meeting ended without a resolution* Stuart and Kiron, "Sirtris Pharmaceuticals," 5.

66 *he promised to call* David Sinclair, phone interview with author, November 13, 2011.

67 *the Kenyon lab researcher Coleen Murphy* Coleen T. Murphy, Steven A. McCarroll, Cornelia I. Bargmann, Andrew Fraser, Ravi S. Kamath, Julie Ahringer, Hao Li, and Cynthia Kenyon, "Genes that Act Downstream of DAF-16 to Influence the Lifespan of *Caenorhabditis elegans*," *Nature* 424, no. 6946 (2003): 277–83.

67 *a Kenyon lab paper in* Cell Natasha Libina, Jen R. Berman, and Cynthia Kenyon, "Tissue-Specific Activities of *C. elegans* DAF-16 in the Regulation of Lifespan," *Cell* 115, no. 4 (2004): 489–502.

67 *you've found an ancient city* "As the Worm Turns," Cynthia Kenyon, interview by Roger Bingham, *The Science Network*, September 23, 2009, accessed June 5, 2010, http://thesciencenetwork.org/programs/diet-aging-and-metabolism/cynthia-kenyon.

67 *big pharma, small pharma* Charles Sawyers, quoted in Marilyn Chase, "Scientists Extend Life Span of Worms by Altering Genes," *Wall Street Journal*, October 24, 2003.

67 *working fourteen hours a day* Nuno Arantes-Oliveira, phone interview with author, February 17, 2004.

67 *how powerful that idea would be* Westphal, quoted in Josh Wolffe, "Christoph Westphal: Finding the Fountain of Youth," *Forbes/Wolfe Emerging*

Tech Report, vol. 8, no. 4 April, 2009 (New York: Forbes, Inc. and Angstrom Publishing), 2.

68 *signed up award-winning researchers* Sirtris company website, accessed January 15, 2006, http://www.sirtrispharma.com.

68 *I asked Christoph many times* David Sinclair, phone interview with author, November 13, 2011.

68 *the unprecedented step of leaving* David Stipp, *The Youth Pill: Scientists at the Brink of an Anti-Aging Revolution* (New York: Current Books, 2010), 228.

68 *a 2004 federal study* National Institute on Aging website, accessed September 22, 2011, http://www.nia.nih.gov/.

68 *fifteen million Americans* Arie Kapteyn, "What Can We Learn from (and about) Global Aging?," Rand Corporation Report WR-741, February 2010 (Santa Monica, CA: Rand Corporation).

69 *a series of self-reinforcing benefits* James Carey and D. S. Judge, "Life Span Extension in Humans Is Self-Reinforcing: A General Theory of Longevity," *Population and Development Review* 27 (2001): 411–36.

69 *controller of a controller* Javier Apfeld, Interview with author, August 16, 2003.

70 *our first presentation to VCs* Michelle Dipp, phone interview with author, June 26, 2008.

70 *different temperatures, different conditions* Matt Kaeberlein, phone interview with author, March 2, 2011.

70 *they're doing exactly what we're doing* Couzin, "Aging Research's Family Feud," 1276.

71 *I never said that* Leonard Guarente, interview with author, April 29, 2004.

71 *everyone in the world could read it* Matt Kaeberlein, phone interview with author, March 2, 2011; M. Kaeberlein, K. T. Kirkland, S. Fields, and B. K. Kennedy, "Sir2-Independent Life Span Extension by Calorie Restriction in Yeast. *PLoS Biology* 2, no. 9 (2004): E296.

71 *peer-reviewed* Journal of Biological Chemistry M. Kaeberlein, T. McDonagh, B. Heltweg, J. Hixon, E. A. Westman, S. D. Caldwell, A. Napper, et al., "Substrate-Specific Activation of Sirtuins by Resveratrol," *Journal of Biological Chemistry* 280:17038–45.

72 *there's a synthesis emerging* David Sinclair, phone interview with author, June 8, 2005.

CHAPTER 7

73 *disparate theories of aging* Steven Austad, phone interview with author, February 10–11, 2005.

73 *maintenance during adult health* Robert Arking, phone interview with author, October 3, 2012.

74 *they look astonishingly human* Monica Driscoll interview by Roger Bingham, "Keystone Symposium on Aging," *The Science Network*, February 10,

2008, accessed June 12, 2010, http://thesciencenetwork.org/programs/keystone-symposia/monica-driscoll.

74 *inhibited the growth of precancerous tumors* Julie M. Pinkston, Delia Garigan, Malene Hansen, and Cynthia Kenyon, "Mutations that Increase the Life Span of *C. elegans* Inhibit Tumor Growth," *Science* 313, no. 5789 (2006): 971–75.

74 *became "celebrity genes"* Steven Austad, "Anti-Aging Therapy: Biological Prospects and Potential Demographic Consequences," panel discussion, American Association for the Advancement of Science, St. Louis, MO, February, 17, 2006.

74 *like a religious cult* Gary Ruvkun interview by Roger Bingham. *The Science Network*, September 10, 2009, accessed August 29, 2012, http://thesciencenetwork.org/programs/diet-aging-and-metabolism/gary-ruvkin.

75 *like Venn diagrams* Marc Tatar, phone interview with author, July 6, 2005.

75 *in the forms of genes and networks* Cynthia Kenyon, "My Adventures with Genes from the Fountain of Youth," In *The Harvey Lectures: Series 100, 2004–2005* (New York: Wiley-Liss, 2006), 63.

75 *a new wave of researcher* Anne Brunet biography, Brunet lab website, accessed November 1, 2011, http://www.brunetlab.stanford.edu.

76 *University of Fribour in Switzerland* Walter Sneader, *Drug Discovery: A History* (New York: Wiley and Sons, 2005), 306.

76 *only a small number of genes* Rick Weindruch, interview with author, September 28, 2009.

76 *tipped the balance towards survival* Linda Partridge, phone interview with author, June 16, 2011.

76 *an Apollo-type government investment* Robert N. Butler, *The Longevity Revolution: The Benefits and Challenges of Living a Long Life* (New York: Public Affairs, 2010), 137.

77 *evaluating its skin-care products* Driscoll interview by Bingham, "Keystone Symposium on Aging."

77 *one of the first of many products* Daniel Maes, quoted in Guy Montague-Jones, "Estee Lauder Reveals Latest Advances in New Anti-Aging Moisturizer," *Cosmetics Today*, October 23, 2008, accessed September 17, 2012, http://www.cosmeticsdesign.com/Formulation-Science/Estee-Lauder-reveals-latest-advances-in-new-anti-aging-moisturizer.

77 *the race to split the atom* Sinclair, quoted in David Stipp, "Researchers Seek Key to AntiAging in Calorie Cutback," *Wall Street Journal*, October 30, 2006.

77 *aging is not programmed* Austad, "Anti-Aging Therapy: Biological Prospects and Potential Demographic Consequences."

77 *$50 million of someone else's money* Marc Tatar, phone interview with author, July 6, 2005.

78 *resveratrol caused no life extension* M. Kaeberlein, T. McDonagh, B. Heltweg, J. Hixon, E. A. Westman, S. D. Caldwell, A. Napper, et al., "Substrate-

Specific Activation of Sirtuins by Resveratrol," *Journal of Biological Chemistry* 280 (2005): 17038–45.

78 *very tricky, very temperature sensitive* David Sinclair, phone interview with author, February 5, 2008.

78 *information laundering operations for the pharmaceutical industry* Richard Horton, "The Dawn of McScience," review of *Science in the Private Interest: Has the Lure of Profits Corrupted Biomedical Research?*, by Sheldon Krimsky, *New York Review of Books*, March 11, 2004, accessed July 8, 2012, http://www.nybooks.com/articles/archives/2004/mar/11/the-dawn-of-mcscience/?pagination=false&printpage=true.

79 *the company struggled for money* Alan Watson, phone interview with author, January 13, 2010.

79 *researchers from the Canadian Medical Expedition* Walter Sneader, *Drug Discovery: A History* (New York: Wiley and Sons, 2005), 307.

80 *turned out to be an essential component* Randy Strong, "Rapamycin Increases Life Span in Mammals," presentation at Harvard Healthy Life Span Conference, September 23, 2009.

80 *after nearly two years* Ibid.

81 *a superintendent of a building* Ziya Tong, "Can We Live Forever?," interview with Cynthia Kenyon, *Nova*, November 12, 2010, accessed July 27, 2011, http://www.pbs.org/wgbh/nova/body/can-we-live-forever.html.

81 *various means of slowing or delaying aging* Arking, *The Biology of Aging*, 377.

82 *cells sense changes in nutrient and hormone signals* Pere Puigserver, presentation at Harvard Healthy Life Span Conference, June 28, 2009.

83 *made me decide to devote my life to this* Michelle Dipp, phone interview with author, July 2, 2008.

83 *amazingly youthful vigor as they aged* David Stipp, *The Youth Pill: Scientists at the Brink of an Anti-Aging Revolution* (New York: Current Books, 2010), 4

83 *the whole field. . . . went gangbusters* Michelle Dipp, phone interview with author, July 2, 2008.

83 *I'm thrilled* Roger Bingham, "Autophagy and Aging," interview with Ana Maria Cuervo, *The Science Network*, accessed July 2, 2012, http://thescience-network.org/programs/keystone-symposia/ana-maria-cuervo.

83 *a whole new science of first principles* Brian Kennedy, phone interview with author, March 15, 2011.

83 *a revolution in our capacity to interfere* Driscoll interview by Bingham, "Keystone Symposium on Aging."

84 *betting a "trifecta of aging"* Andrew Dillin and Suzanne Wolff, "The Trifecta of Aging in *C. elegans*," *Experimental Gerontology* 41, no. 10 (2006): 894–903.

CHAPTER 8

87 *life spans could lengthen or shorten* David Reznick, phone interview with author, May 1, 2011; D. Reznick, M. Bryant, and D. Holmes, "The Evolution

of Senescence and Post-Reproductive Lifespan in Guppies," *PLoS Biology* 4, no. 1 (2005): e7.

88 *pattern of growing old* David Reznick, phone interview with author, August 30, 2006.

88 *there's more to life than making babies* "Guppies have Menopause, Too," Science on MSNBC, updated December 30, 2005, accessed March 22, 2012, http://www.msnbc.msn.com/id/10643970/Reznick.

88 *a rich breeding ground for old animals* "The Lifespan of Animals," last modified November, 2011, accessed December 2, 2011, http://www.newton.dep.anl.gov/natbltn/400–499/nb486.htm; Leonard Hayflick, phone interview with author, November 8, 2011.

88 *a family of chemicals vying with each other* David Reznick, phone interview with author, August 30, 2006.

88 *the lab researchers hate this* Steven Austad , Steven Austad, "Anti-Aging Therapy: Biological Prospects and Potential Demographic Consequences," panel presentation, American Association for the Advancement of Science Panel Presentation, St. Louis, MO, February 17, 2006.

89 *exactly what we were trying to emulate* Gary Ruvkun, quoted in "Diet, Aging and Metabolism," *The Science Network*, June 8, 2010, accessed July 6, 2012, http://thesciencenetwork.org/programs/diet-aging-and-metabolism/gary-ruvkun.

89 *seven bodybuilding genes align* Israel Rosenfield and Edward Ziff, "Evolving Evolution," *New York Review of Books*, May 11, 2006, 12–16.

89 *building flies for 200 million years* Sydney Brenner, "Francis Crick in Paradiso," Loose Ends, *Current Biology* 6 (1996): 1202.

90 *backlash against the post-genome determinism* Linda and Edward McCabe, *DNA: The Promise and Peril* (Los Angeles: University of California Press, 2007), 80–200; 112.

91 *the long-lived animals that evolution has produced* Steven Austad, phone interview with author, January 25, 2012.

91 *I don't have time in my life anymore* David Reznick, phone interview with author, May 1, 2012.

91 *elderly females taught effective response* Cynthia Moss, on "Echo: An Elephant to Remember," *Nature*, PBS, October 17, 2010, accessed August 29, 2012, http://www.pbs.org/wnet/nature/episodes/echo-an-elephant-to-remember/video-full-episode/5920/.

91 *recognized earliest the warnings* James Jackson, "Surviving Natural Disaster," in *Survival of the Human Race* ed. Emily Shuckburgh (Cambridge: Cambridge University Press, 2008), http://www.globalaging.org/armedconflict/countryreports/asiapacific/tsunami.conference.htm#panel.

92 *looked much like that of the long-lived lab mutants* Richard Miller, phone interview with author, July 6, 2010.

92 *secret seemed to be stress resistance* "Richard A. Miller, MD, PhD," last updated December 22, 2007, accessed July 3, 2010, http://www-personal.umich.edu/~millerr/RAM_home_page.htm.

92 *one of the only known animals* "Carol Vleck: Research," last updated 2007, accessed November 3, 2011, http://www.public.iastate.edu/~cvleck/research.html.

92 *a haunting set of photographs* Caleb Finch and Robert Ricklefs, *Aging: A Natural History* (New York: Scientific American Library, 1995), 8.

93 *what naked mole rats had* Rochelle Buffenstein, phone interview with author, September 8, 2011.

93 *single gene pathway related to a single characteristic* David Reznick, phone interview with author, May 1, 2012.

93 *truly a physiological shift* Cynthia Kenyon, phone interview by author, February 4, 2011.

93 *I absolutely believe that aging is programmed* Rich Morimoto, interview with author, December 15, 2011.

93 *it's just very complicated and intricate* Joy Alcedo, interview with author, July 22, 2002.

94 *a ticking time bomb* Eline Slagboom, phone interview with author, August 2, 2011.

94 *spent twenty years answering that question* Manja Shoenmaker, Anton J. M. de Craen, Paul H. E. M. de Meijer, Marian Beekman, Gerard J. Blauw, P. Eline Slagboom, and Rudy G. J. Westendorp, "Evidence of Genetic Enrichment for Exceptional Survival Using a Family Approach: The Leiden Longevity Study," *European Journal of Human Genetics* 14 (2006): 79–84.

95 *but it is a system, not one gene* Eline Slagboom, phone interview with author, August 2, 2011.

96 *the most consistent validated gene* Nir Barzilai, phone interview with author, February 17, 2012.

96 *the question was so urgent it funded a study* S. Jay Olshansky, interview with author, October 10, 2011.

97 *the next round of financing* Toby Stuart and David Kiron, "Sirtris Pharmaceuticals: Living Healthier, Longer," Harvard Business School Case Study N9–808–112 (March 18, 2008), 8.

97 *remained skeptical about sirtuins' role* Charlie Schmidt, "GSK/Sirtris Compounds Dogged by Assay Artifacts," *Nature Biotechnology* 28 (2010): 185–186.

97 *polyphenols typically show a lot of different effects* Derek Lowe, phone interview with author, December 16, 2011.

97 *the worldwide corporation was restructuring* Toby Stuart and James Webb, "GSK's Acquisition of Sirtris: Independence or Integration?," Harvard Business School Case Study N9–808–112 (April 13, 2009), 3.

97 *the resveratrol study had more going for it* David Stipp, *The Youth Pill: Scientist at the Brink of an Anti-Aging Revolution* (New York: Current Books, 2010), 9.

98 *Elixir prepared to go public* Alan Watson, phone interview with author, June 21, 2010.

99 *what Kenyon and others are doing* David Reznick, phone interview with author, May 1, 2012.

CHAPTER 9

100 *as if he should have worn a "papal ring"* David Stipp, *The Youth Pill: The Rise of Biogerontology and the Race to Slow Aging* (New York: Current, 2010), 234.

100 *the company had made the cover of* Fortune David Stipp, "Drink Wine and Live Longer," *Fortune*, February 23, 2007, 22–25.

100 *anti-aging message is very powerful* Christoph Westphal, quoted in Toby Stuart and David Kiron, "Sirtris Pharmaceuticals: Living Healthier, Longer," Harvard Business School Case Study N9–808–112 (March 18, 2008), 6–7.

100 *the giant pharmaceutical manufacturer GlaxoSmithKline* GlaxoSmithKline website, accessed June 18, 2012, http://www.gsk.com/about/ataglance.htm.

100 *required several levels of approval* Toby Stuart and James Webb, "GSK's Acquisition of Sirtris: Independence or Integration?," Harvard Business School Case Study N9–808–112 (April 13, 2009), 9.

101 *a "potentially transformative science"* Douglas Sipp, "GSK Moves on Sirtris," *Nature Biotechnology* 26, no. 595 (2008), doi:10.1038/nbt0608-595.

101 *if we are wrong, we might get nothing* Andrew Witty, quoted in Stuart and Webb, "GSK's Acquisition of Sirtris," 7

101 *biotech and business media applauded* Sipp, "GSK Moves On Sirtris," 26.

102 *you could not buy better advertising* Michelle Dipp, phone interview with author, July 2, 2008.

102 *downgrade the expectations* Christoph Westphal, quoted in Josh Wolfe, "Christoph Westphal: Finding the Fountain of Youth," *Forbes/Wolfe Emerging Tech Report*, April 2009, 3.

102 *hard for the field to advance* Cynthia Kenyon, interview with author, September 28, 2009.

102 *punished for explaining your science correctly* Peter Distefano, phone interview with author, March 1, 2011.

103 *first major pharmaceutical company to make a major bet* Paul Glenn, introduction to the Harvard Healthy Life Span Conference, Boston, MA, June 23, 2008.

103 *Sinclair joined the paid board of Shaklee* Shaklee Corporation website, accessed December 1, 2011, http://www.shaklee.com/index.shtml.

104 *claiming it had misused his quotes* Keith J. Weinstein, "Harvard Anti-Aging Researcher Quits Shaklee Advisory Board," *Wall Street Journal*, December 26, 2008, accessed December 2, 2011, http://online.wsj.com/article/SB123025446150734561.html.

104 *every Shaklee use of his name* "Shaklee Loses Major Endorser," MarketWave

.com, December 30, 2008, accessed December 11, 2011, http://www.marketwaveinc.com/viewalert.asp?id=107.

104 *public fascination skyrocketed* Jon Hamilton, "Scientists Develop Life-Extending Compounds," NPR, updated November 29, 2007, accessed November 3, 2011, http://www.npr.org/templates/story/story.php?storyId=16727282; "Wine Rx," *60 Minutes*, aired January 25, 2009, CBS, accessed December 12, 2011, http://www.cbsnews.com/video/watch/?id=5037314n&tag=mncol;lst;1.

104 *so is Dr. Oz saying they're good or not* "The Truth About Oprah, Dr. Oz, Acai, Resveratrol and Colon-Cleanse," Oprah.com, accessed July 5, 2012, http://www.oprah.com/health/The-Truth-About-Oprah-Dr-Oz-Acai-Resveratrol-and-Colon-Cleanse.

104 *cited as much as 72 percent more* Daniel R. Phillips, et al. "Importance of the Lay Press in the Transmission of Medical Knowledge to the Scientific Community," *New England Journal of Medicine* 325 (1991): 1180–83.

104 *if you wanted a story to sell* George Vlasuk, phone interview with author, December 5, 2011.

104 *a big weight on my shoulders* David Sinclair, phone interview with author, September 28, 2011.

105 *foot to the pedal* Andy Dillin, interview with author, December 11, 2010.

105 *garbage can that processes or recycles* Nir Barzilai and Ana Maria Cuervo interview by Roger Bingham, *The Science Network*, January 20, 2011, accessed July 18, 2011, http://thesciencenetwork.org/programs/new-york-city-january-2011/nir-barzilai-and-ana-maria-cuervo-1.

105 *the connections among these overlapping networks* Anne Brunet, phone interview with author, November 16, 2011.

105 *recognized the biology of aging as a science* "The Nobel Prize in Physiology or Medicine 2009," NobelPrize.org, accessed December 4, 2011, http://www.nobelprize.org/nobel_prizes/medicine/laureates/2009/.

106 *Kenyon's lab sought compounds that could affect* Cynthia Kenyon, phone interview, December 14, 2011.

106 *Ruvkun's lab found that the FOXO gene triggered* S. P. Curran, X. Wu, C. G. Riedel, and G. A. Ruvkun, "Soma-to-Germline Transformation in Long-Lived *Caenorhabditis elegans* Mutants," *Nature*. 459, no. 7250 (2009): 1079–84.

106 *studied a group of small-sized Ecuadorians* Valter Longo, phone interview with author, September 22, 2011.

106 *monitored his rhesus monkeys on caloric restriction* Richard Weindruch, in discussion with the author, Harvard Healthy Life Span Conference, Boston, MA, September 28, 2009.

106 *no longer a question of if, but when* Jeffrey Flier, Introduction to Harvard Healthy Life Span Conference, Boston, MA, September 28, 2009.

107 *the problem with the original* Science *and* Nature *papers* Matt Kaeberlein, phone interview with author, February 28, 2011.

107 *a 2011 study by two medical journal editors* Paul Jump, "Retractions on the Rise" *Times Higher Education*, August 25, 2011, accessed October 27, 2011, http://www.insidehighered.com/news/2011/08/25/study_finds_increase_in_number_of_journal_articles_retracted.

107 *while we were doing this [fly] work* Linda Partridge, phone interview with author, June 16, 2011.

108 *if they did a standard genetic outcross* David Gems, phone interview with author, June 9, 2010.

108 *unwittingly allowed a second gene* Heidi Ledford, "Much Ado about Ageing," *Nature* 464 (2010): 480–81.

108 *Amgen pharmaceutical scientists announced* Beher et al., "Resveratrol Is Not a Direct Activator of SIRT1 Enzyme Activity," *Chemical and Biological Drug Design* 74 (2009): 619–24.

108 *a month later Pfizer pharmaceutical researchers* M. Pacholec et al., "SRT 1720, SRT2183, SRT 1460 and Resveratrol Are Not Direct Activators of SIRT1," *Journal of Biological Chemistry* 285, no. 11 (2010): 8340–51.

108 *reported results were "a complete artifact"* Peter Distefano, phone interview with author, March 2, 2011.

108 *instead they hit multiple targets* Derek Lowe, "Sirtris' Compounds: Everyone Agrees?," In the Pipeline (blog), Corante.com, accessed July 14, 2010, http://pipeline.corante.com/archives/2010/04/28/sirtriss_compounds_everyone_agrees.php.

108 *"debate in the academic world"* Kevin Davies, "Christoph Westphal Dishes on Aging and Pharmageddon," Bio-IT World, April 26, 2010, accessed December 1, 2010, http://www.bio-itworld.com/news/04/26/10/Christoph-Westphal-on-aging-pharmageddon.html.

108 *not at all surprised there's some controversy* Andrew Witty, quoted in John Caroll, "Glaxo Spurns New Research Raising Doubts about Resveratrol," *FierceBiotech*, January 26, 2010, accessed July 5, 2012, http://www.fiercebiotech.com/story/glaxo-spurns-new-research-raising-doubts-about-resveratrol/2010-01-26.

109 *there's not much overlap* Derek Lowe, phone interview with author, December 16, 2011.

109 *the assay is worthless* Lowe, "GSK's Response to the Sirtuin Critics."

109 *I'm actually quite excited about it* Derek Lowe, phone interview with author, December 16, 2011.

109 *scientists are supposed to remain impartial* Cynthia Kenyon, e-mail to author, February 24, 2010.

110 *moved to prevent Westphal and Dipp from doing so* Adam Feurerstein, "Glaxo Slaps Former Sirtris Execs," *The Street*, accessed August 12, 2010, http://www.thestreet.com/story/10835296/1/glaxo-slaps-former-sirtris-execs.html.

110 *not as developed as it should have been* George Vlasuk, phone interview with author, December 5, 2011.

110 *the* SIR *bubble* David Gems, phone interview with author, June 9, 2010.

111 *the next day the headline reads* David Sinclair, phone interview with author, November 23, 2011.

111 *it's what science does* Linda Partridge, phone interview with author, June 16, 2011.

111 *there's a lot of potential there* Derek Lowe, phone interview with author, December 16, 2011.

111 *China invested heavily* "Brief Introduction to Guangdong Medical College," Guangdong Medical College website, accessed October 6, 2011, http://www.gdmc.edu.cn/english/sco/egdyjj.htm.

111 *but moving incredibly fast* Matt Kaeberlein, phone interview with author, February 22, 2011.

CHAPTER 10

113 *I'm into promoting youth* Andrew Dillin, interview with author, December 7, 2010.

113 *a new generation of researchers* Murphy Lab at Princeton University website, accessed October 27, 2011, http://molbio.princeton.edu/faculty/molbio-faculty/120-murphy; Brunet Lab at Stanford University website, accessed November 15, 2011, http://www.stanford.edu/group/brunet/; "Heidi A Tissenbaum, PhD," University of Massachusetts website, accessed September 20, 2011, http://profiles.umassmed.edu/profiles/ProfileDetails.aspx?From=SE&Person=658; "Marlene Hansen, PhD," Sanford Burnham Medical Institute, accessed December 27, 2011, http://www.sanfordburnham.org/research_and_faculty/faculty_search/hansen_m_phd.aspx.

113 *we've become Rapa-land* David Harrison, quoted in Gary Taubes, "Timeless and Trendy Effort to Find—or Create—the Fountain of Youth," *Discover*, October 2010, accessed April 22, 2012, http://discovermagazine.com/2010/oct/12-timeless-trendy-effort-find-create-fountain-youth.

114 *ask any gerontologist* Judith Campisi, phone interview with author, June 14, 2011.

114 *disease-fighting genes* Gordon Lithgow, phone interview with author, September 8, 2011.

115 *all of my work is on protein folding* Andrew Dillin, interview with author, December 7, 2010.

115 *future directions in biomedical research* I. Benjamin et al., "Stress (Heat Shock) Proteins: Molecular Chaperones in Cardiovascular Biology and Disease," *Aging*; Rick Morimoto, interview with author, December 15, 2011.

116 *the tissues where TOR translation takes place* Kristan Steffen, interview with author, December 7, 2010.

116 *it would have had a magic effect* Andrew Dillin, interview with author, December 7, 2010.

116 *how many hours a week she worked* Anne Brunet, phone interview with author, November 16, 2011.

117 *so many connections between music and science* Erik Kempernick, interview with author, December 7, 2010.

117 *an astonishing 247 known or suspected longevity genes* Gordon Lithgow, phone interview with author, September 8, 2011.

118 *it can attach to a specific spot* Brent Stockwell, *The Science and Stories Behind the Quest for the Next Generation of Medicines* (New York: Columbia University Press, 2011), 101.

118 *SRT2104 was being tested in phase 2* You can visit the ClinicalTrials website at http://clinicaltrials.gov and search for results.

119 *sirtuins' roles as metabolic regulators* George Vlasuk, phone interview with author, December 2, 2011.

119 *an early understanding of the biology* George Vlasuk, phone interview with author, December 2, 2011.

119 *was testing metformin for type 2 diabetes* ClinicalTrials.gov, a service to the U.S. National Institutes of Health," accessed November 29, 2011, http://www.clinicaltrials.gov.

119 *some of the most successful drugs* Morton A. Meyers, *Happy Accidents: Serendipity in Modern Medical Breakthroughs* (Chicago: Arcade Publishing, 2007), xiii.

120 *the biotech dream continued* David Thomas, "Venture Capital Increases in 2011, but . . . ," Biotech Now, accessed June 18, 2012, http://www.biotech-now.org/business-and-investments/inside-bio-ia/2012/01/vc2011.

120 *studying scrawny animals, "tubes with jaws"* Rochelle Buffenstein, phone interview with author, September 8, 2011.

121 *resisted the chemicals that caused free radical damage* Manlio Vinciguerra, "Hormones, Reproduction and Disease in the Longest-Lived Rodent: The Naked Mole Rat," *Endocrinology Studies* 1, no. 1 (2011): e4, 14–16.

121 *longevity itself was not adaptive* Rochelle Buffenstein, phone interview with author, September 8, 2011.

121 *cytoprotective, detoxifying mechanisms or agents* Ibid.

121 *observed a contradiction to the Hayflick limit* J. Campisi, "Replicative Senescense: An Old Lives' Tale?" *Cell* 23;84, no. 4 (1996): 497–500.

122 *a basic cellular mechanism linking cancers* Judith Campisi, phone interview with author, June 14, 2011.

122 *mice with INK-ATTAC remained healthy* Brandon Keim, "Cell-Aging Hack Opens Longevity Research Frontier," *Wired*, November 2, 2011, accessed January 3, 2012, http:// www.wired.com/wiredscience/2011/11/cellular -senescence.

122 *fat and muscle tissue damage* Darren J. Baker, Tobias Wijshake, Tamar Tchkonia, Nathan K. LeBrasseur, Bennet G. Childs, Bart van de Sluis, James L. Kirkland, and Jan M. van Deursen, "Clearance of p16ink4a

-Positive Senescent Cells Delays Ageing-Associated Disorders," *Nature* 479 (2011): 232–36.

123 *there are processes driving aging* Judith Campisi, quoted in Nicholas Wade, "Purging Cells in Mice is Found to Combat Aging Ills," *New York Times*, November 15, 2011, accessed November 18, 2011, http://www.nytimes.com/2011/11/03/science/senescent-cells-hasten-aging-but-can-be-purged-mouse-study-suggests.html.

123 *a mouse finding on the front page* Nicholas Wade, phone interview with author, November 3, 2011.

123 *venture capital firm approached them* Judith Campisi, phone interview with author, August 8, 2012.

123 *age of epigenetics has arrived* John Cloud, "Why Your DNA Isn't Your Destiny," *Time*, January 26, 2010, accessed May 13, 2012, http://www.time.com/time/magazine/article/0,9171,1952313,00.html.

123 *some percentage are diseases of epimutation* "Epigenetics: Supplemental Materials," University of Utah Genetics Science Learning Center, accessed January 4, 2012, http://teach.genetics.utah.edu/content/epigenetics/.

124 *epimutations happen in one of three ways* "DNA Structure and Protein Expression," Epizyme, accessed January 8, 2012, http://www.epizyme.com/epigenetics/dna-structure-protein-expression.asp.

124 *chronic age-related pathological states* Trygve O. Tollesbol, ed., *Epigenetics of Aging* (Oxford: Oxford University Press, 2009), 17.

124 *the pattern of such epimutations* Richard C. Francis, *Epigenetics: The Ultimate Mystery of Inheritance* (New York: Norton, 2011), 4

124 *prenatal studies of obese mice* Thomas Rando, "Epigenetics and Aging," *Experimental Gerontology* 45, no. 4 (2010): 253–54.

125 *increase in oncogene expression* Holly Brown-Borg, phone interview with author, September 5, 2012.

125 *it seems very fatalistic* David Reznick, phone interview with author, May 1, 2011.

125 *thousands of epigenetic marks* Holly Brown-Borg, phone interview with author, September 28, 2011.

125 *a bit mushy* Cynthia Kenyon, phone interview with author, July 12, 2011.

126 *trying to get the cancer cells* Jean Pierre Issa, quoted in "Epigenetics," *Nova*, July 1, 2007, accessed November 12, 2011, http://www.pbs.org/wgbh/nova/body/epigenetics.html.

126 *the FDA approved decitabine* University of Texas MD Anderson Cancer Center website, accessed December 13, 2011, http://www.mdanderson.org/education-and-research/research-at-md-anderson/early-detection-and-treatment/research-programs/spores/leukemia-spore/investigators-and-staff/index.html.

126 *much of the new aging epigenetics* D. Sinclair et al., "SIRT1 Redistribution on Chromatin Promotes Genomic Stability but Alters Gene Expression During Aging," *Cell* 135 (2008): 907–11.

126 *a nice youthful pattern of DNA repair* David Sinclair, phone interview with author, November 23, 2011.

126 *studied sirtuins' epigenetic effects on Alzheimer's* "J. Lawrence Marsh," Faculty Profile System, University of California, Irvine, accessed December 27, 2011, http://www.faculty.uci.edu/profile.cfm?faculty_id=2721.

126 *studied sirtuins' epigenetic effects on cancer* "Katrin Chua," Community Academic Profiles, Stanford School of Medicine, accessed December 27, 2011, http://med.stanford.edu/profiles/Katrin_Chua/.

126 *sirtuins' role in DNA repair and metabolism* Mostoslavsky Lab website, Massachusetts General Hospital Cancer Center, accessed December 1, 2011, http://www.massgeneral.org/cancer/research/researchlab.aspx?id=1185.

126 *expanded their epigenetics programs* CellCentric website, accessed December 1, 2011, http://www.cellcentric.com/.

126 *the compounds that control the chromatin* Brad Johnson, phone interview with author, June 22, 2011.

127 *largest epidemiological study of natural resveratrol's healthful effects* Howard Lovy, "Danes Begin Biggest-Ever Resveratrol Study," *Fierce Biomarkers*, February 1, 2011, accessed June 23, 2011, http://www.fiercebiomarkers.com/story/danes-begin-biggest-ever-resveratrol-study/2011-02-01. Accessed June 23, 2011

127 *obese mice fed resveratrol* Nicholas Wade, "Longer Lives for Obese Mice, With Hope for Humans of All Sizes," *New York Times*, August 18, 2011.

CHAPTER 11

128 *other countries were closing their borders* Thai/Cambodia Border Refugee Camps 1975–1999 Information and Documentation website, last updated February 13, 2011, accessed January 15, 2012, http://www.websitesrcg.com/border/documents/Cambodian-Emergency-Refugee-Health-Care-1979-1980.pdf.

128 *moved to train in endocrinology* Nir Barzilai, phone interview by author, November 2, 2011.

128 *a genetic difference in human aging* Ibid.

129 *Albert Einstein College of Medicine's Institute of Aging Research* The Longevity Genes Project, accessed January 15, 2012, http://www.einstein.yu.edu/aging/longevity-genes-project.aspx?utm_source=ein-cpr&utm_medium=redirect&utm_campaign=agingproj.

129 *the competition featured Japanese Hawaiians* The Okinawa Centenarian Study, accessed November 10, 2011, http://www.okicent.org/team.html; "Friederike Flachsbart," Resolve website, accessed January 9, 2012, http://resolve.punkt-international.eu/index.php?id=214; "Eline Slagboom: Faculty Member in Aging," Faculty of 1,000, accessed August 10, 2011, http://f1000.com/thefaculty/member/1677635567244806.

129 *the Archon Genomics X Prize* Archon Genomics X Prize, accessed January 10, 2012, http://genomics.xprize.org/.

130 *that's the elephant in the room* Nir Barzilai, phone interview with author, November 2, 2011.

130 *one of a number of competing humanoid species* Nicholas Wade, *Before the Dawn: Recovering the Lost History of Our Ancestors* (New York: Random House, 2011), 12.

130 *maintained a storied culture* Wikipedia, "Ashkenazi Jews," accessed January 9, 2012, http://en.wikipedia.org/wiki/Ashkenazi_Jews.

131 *showed that they shared a mutation* Y. Suh, G. Atzmon, M. Cho, D. Hwang, B. Liu, N. Barzilai, and P. Cohen, "Functionally Significant Insulin-Like Growth Factor-I Receptor Mutations in Centenarians," *Proceedings of the National Academy of Sciences* 105 (2008): 3438–42.

131 *wanted to learn their secrets* Bradley Willcox, phone interview with author, November 11, 2011.

131 *they [the participants] hit the jackpot* Ibid.

132 *turn on the FOXO transcription factor* Bradley Willcox et.al., "FOXO3a Strongly Associated with Human Longevity," *Proceedings of the National Academy of Sciences* 105 (2008): 13987–92.

132 *the human version of the worm gene* Eline Slagboom, phone interview with author, August 2, 2011.

132 *may provide a potential bridge* Nir Barzilai, e-mail to author, April 29, 2012.

133 *we had an inadequate language* Ronald M. Green, *Babies By Design: Ethics of Genetic Choice* (New Haven, CT: Yale University Press, 2007), 7.

133 *the tools are a little primitive* Bradley Willcox, phone interview with author, November 14, 2011.

134 *not a unique four letter sequence* Barry Barnes and John Dupre, *Genomes and What to Make of Them* (Chicago: University of Chicago Press, 2011).

134 *two respected human longevity gene researchers* Carolyn Y. Johnson, "BU Researchers Retract Genetic Study of Extreme Longevity," July 22, 2011, http://www.boston.com/news/science/articles/2011/07/22/bu_researchers_retract_genetic_study_of_extreme_longevity/?s_campaign=8315.

134 *recapped the sirtuin mistakes* Jennifer Couzin-Frankel, "The Sirtuin Story Unravels," *Science*, December 2, 2011, 1197.

134 *public pronouncements got out of hand* Matt Kaeberlein, phone interview with author, December 21, 2011.

134 *going to get even more confusing* Bradley Willcox, phone interview with author, November 14, 2011.

135 *have something to look forward to* The Longevity Genes Project.

135 *untapped area of aging biology* Nir Barzilai, phone interview with author, November 2, 2011.

135 *centenarians do not die from heart disease* Bradley Willcox, phone interview with author, November 14, 2011.

136 *the moment when they stop reproducing* Cindy Voisine, interview with author, December 15, 2011.

136 *longevity is about the edges* Rick Morimoto, interview with author, December 15, 2011.

136 *found out something about protein* Kurt Vonnegut, *Cat's Cradle* (New York: Dell, 1998), 24–25.

137 *going to be real game-changers* Coleen Murphy, phone interview with author, October 28, 2011.

137 *the voice on the phone* Nir Barzilai, phone interview with author, November 2, 2011.

138 *can you figure that one out* Ibid.

139 *the way it responds to the environment* Coleen Murphy, phone interview with author, October 28, 2011.

139 *not what we've been told* Cynthia Kenyon, at Harvard Healthy Life Span Symposium, Boston, MA, September 28, 2009.

CHAPTER 12

141 *the world's 80-year-olds* United Nations, "Population Aging and Development 2009," accessed January 10, 2012, http://www.un.org/esa/population/publications/ageing/ageing2009chart.pdf.

141 *the number of people age 65 or older* United Nations, "Ageing: Social Policy and Development Division," accessed June 19, 2012, http://social.un.org/index/Ageing.aspx.

141 *the biggest rise in the elderly's numbers* "Unprecedented Global Aging Examined in New Census Bureau Report Commissioned by the National Institute on Aging," press release, July 20, 2009, http://www.census.gov.

141 *now makes a powerhouse* American Association of Retired Persons, accessed July 21, 2011, http://www.aarp.com.

142 *nonprofits like the American Federation for Aging Research* American Federation for Aging Research, accessed September 9, 2011, http://www.afar.org/; Gerontological Society of America, accessed December 11, 2011, http://www.geron.org; Gray Panthers, accessed January 11, 2012, http://www.graypanthers.org/; Older Women's League: The Voice of Midlife and Older Women," accessed January 12, 2011, http://www.owl-national.org/; International Longevity Centre–UK, accessed January 13, 2011, http://www.ilcuk.org.uk/.

142 *long-range gloom or gluttony* Mary Furlong, *Turning Silver into Gold How to Profit in the New Boomer Marketplace* (New York: FT Press, 2011).

142 *newest initiative is the MacArthur Foundation Network* The John D. and Catherine T. MacArthur Foundation, accessed August 31, 2012, http://www.agingsocietynetwork.org.

142 *the longevity revolution is a fragile one* S. Jay Olshansky, Toni Antonucci, Lisa Berkman, Robert H. Binstock, Axel Boersch-Supan, John T. Cacioppo, Bruce A. Carnes, et al., "Differences in Life Expectancy Due to Race and Educational Differences Are Widening, and Many May Not Catch Up," Bethesda, MD, Project Hope, *Health Affairs* 31, no. 8 (2012):1803–13.

143 *youth is in your genes* Advertisement for Lancôme Génifique, *Elle*, October 2011, 111.

143 *it is now 67.6* Mark L. Haas, "Pax Americana Geriatrica," *Miller-McCune*, July 14, 2008, http://www.miller-mccune.com/culture-society/pax-americana-geriatrica-4416/.

143 *rose from 5 to 7 billion people* "World Population Boom," *AARP Bulletin*, October 2011, 34.

144 *number of centenarians is expected to increase* "Ageing in the 21st Century: A Celebration and a Challenge," Gerontological Society of America, accessed October 10, 2012, http://unfpa.org/ageingreport.

144 *bankrupting Social Security in twenty-five years* U.S. Social Security Administration, "A summary of the 2012 Annual Reports," modified June 4, 2012, accessed June 12, 2012, http://www.socialsecurity.gov/OACT/TRSUM/index.html.

144 *four Americans work for every retired person* Kelly Holder and Sandra Clark, "Working Beyond Retirement Age," U.S. Census Bureau PowerPoint presentation, American Sociological Conference, August 2, 2008, accessed July 6, 2012, http://www.census.gov/hhes/www/laborfor/Working-Beyond-Retirement-Age.pdf.

144 *people older than sixty will outnumber* Hyeoun-Ae Park et al., *Consumer Centered Consumer Supported Care for Healthy People: Proceedings of NI2006* (IOS Press, 2006), 4.

144 *workforces are expected to shrink* "Japan's Rapidly Shrinking Workforce," Japan Investor, modified January 31, 2012, accessed May 10, 2012, http://www.japaninvestor.net/2012/01/japans-rapidly-shrinking-workforce.html; "Russia's Workforce is Ageing," New Europe Online, modified March 25, 2012, accessed May 10, 2012, http://www.neurope.eu/article/russia-s-workforce-ageing.

144 *Germany's by one-fifth* "Germany Faces Up to Ageing Workforce," *Guardian*, modified March 27, 2011, accessed May 17, 2012, http://www.guardian.co.uk/world/2011/mar/17/new-europe-germany-retirement-pensions-exports.

144 *forthcoming advances in the biomedical sciences* S. J. Olshansky et al., "Aging in America in the Twenty-first Century: Demographic Forecasts from the MacArthur Foundation Research Network on an Aging Society," *Milbank Quarterly* 87, no. 4 (2009), 846.

145 *two countries, identical in all respects* David Bloom and David Canning, "The Health and Wealth of Nations, *Science*, February 2000, 1207–9.

145 *a fifty-seven-year-old health insurance auditor* Roxanne Aune, phone interview with author, December 20, 2011.

145 *discrimination, subtle and unprovable* Roxanne Aune, quoted in Sally Abrahams, "The Caregiver's Dilemma," *AARP Bulletin*, September, 2011, 10–12, accessed July 7, 2012, http://www.sallyabrahms.com/articles/detail.asp?content=94&category=3.

145 *national gross domestic product goes to health care* Tom Donley, "The Obama Health Plan" (DePaul University, talk at Grace Lutheran Church River Forest, IL, February 8, 2010).

146 *one worker will be supporting six people* Ted Fishman, *Age Shock: How the Aging Society Will Pit Boss Against Worker, Generation against Generation* (New York: Crown, 2010), 297.

146 *age discrimination pays* Ibid., 7.

146 *number of Americans over the age of fifty* Mark Watson and James Stock. "Gray Nation, The Very Real Dangers of an Aging America," *Atlantic*, accessed April 29, 2012, http://www.theatlantic.com/business/archive/2012/03/gray-nation,-the-very-real-dangers-of-an-aging-america/254937.

146 *the probability of being poor* "4 in 10 Americans over 60 will experience poverty: AARP Study," *Senior Journal*, modified May 23, 2001, accessed October 23, 2011, http://seniorjournal.com/NEWS/Features/05-23-1AARPStudy.

147 *medical bankruptcies today* Christine Dugas, "Bankruptcies Rising among Seniors," *USA Today*, June 20, 2008, accessed July 8, 2012, http://www.usatoday.com/money/perfi/retirement/2008–06–16-bankruptcy-seniors_N.htm.

147 *6 percent of the gross domestic product* Donley, "The Obama Health Plan."

147 *people are caring for an aging adult* Strength for Caring, "Family Caregiving in America: Facts at a Glance," accessed July 6, 2012, http://www.strengthforcaring.com/util/press/facts/facts-at-a-glance.html.

147 *National Alliance for Caregiving study* "National Alliance for Caregiving," December 21, 2011; Abrahams, "The Caregiver's Dilemma."

147 *complain about health problems* Susan Jacoby, *Never Say Die: The Myth and Marketing of the New Old Age* (New York: Harper, 2011), xii.

148 *a* Time *cover featured "Greys on the Go"* Robert Binstock, *Aging Nation: The Politics and Economics of Growing Older in America* (Baltimore, MD: Johns Hopkins University Press, 1998), 8.

148 *there are the young old, and the old old* S. Jay Olshansky, interview with author, October 17, 2011.

149 *seeing the power of the new lab discoveries* S. Jay Olshansky, Ph.D., accessed October 16, 2011, http://www.cade.uic.edu/sphapps/faculty_profile/sphFacultyInd.asp?i=sjayo&d=.

149 *if you're in a manual trade* S. Jay Olshansky, interview with author, October 17, 2011.

150 *an astonishing, beneficial fall* "Falling Fertility: Astonishing Falls in the Fertility Rate Are Bringing with Them Big Benefits," *Economist*, October 29, 2009, accessed February 1, 2012, http://www.economist.com/node/14744915.

150 *world gross domestic product increased nineteenfold* Bradford De Long, "Estimating World GDP, One Million B.C. to the Present," accessed July 8, 2012, http://delong.typepad.com/print/20061012_LRWGDP.pdf.

151 *the new Medicare prescription plan* Paula Span, "Changes to Medicare," *New York Times*, February 9, 2011, accessed June 29, 2012, http://newoldage.blogs.nytimes.com/2011/02/09/changes-to-medicare/?scp=3&sq=Medicare+Prevention&st=nyt.000 *a number of reports* U.S. Senate Special Committee on Aging, *Social Security Modernization: Options to Address Solvency and Benefit Adequacy* (Washington, DC: US Government Printing Office, 2010), http://aging.senate.gov/letters/ssreport2010.pdf.

151 *straightforward steps to maintaining solvency* U.S. Social Security Administration, "Proposals Addressing Trust Fund Solvency," June 21, 2012, http://www.ssa.gov/OACT/solvency/; Mary Beth Franklin, "How to Fix Social Security," *Kiplinger*, September 2011, accessed June 23, 2012, http://www.kiplinger.com/columns/retirement/archives/how-to-fix-social-security.html.

151 *German economist Axel Boersch-Supan* National Bureau of Economic Research, "Axel Boersch-Supan," accessed October 16, 2011, http://www.nber.org/people/axel_boersch-supan.

152 *research into the biology of aging* Robert Arking, *The Biology of Aging: Observations and Principles* (New York: Oxford University Press, 2006), vi.

152 *argued we should not accept chronic disease* Aubrey de Grey, *Ending Aging: The Rejuvenation Breakthroughs That Could Reverse Human Aging in Our Lifetime* (New York: St. Martin's Press, 1963).

152 *shift in the fundament of human expectation* Ray Kurzweil, *The Singularity Is Near: When Humans Transcend Biology* (New York: Penguin, 2005).

153 *that's not science* Jay Olshansky, interview with author, October 17, 2011.

154 *if older people feel* Kathleen Woodward, *Aging and Its Discontents: Freud and Other Fictions (Theories of Contemporary Culture* (Bloomington: Indiana University Press, 1991), 7.

154 *to keep Social Security solvent* Jane Bryant Quinn, "Myths and Truths about Social Security," *AARP Bulletin*, November 2011, accessed June 2, 2012, http://www.aarp.org/work/social-security/info-11–2011/5-social-security-myths.2.html.

154 *similar adjustments may be considered* Jonathan Peterson, "Time for a Tune-Up," *AARP Bulletin*, June 2012, 12; Patricia Barry, "Re-Tooling Medicare," *AARP Bulletin*, June 2012, 8

155 *Valter Longo was invited* Valter Longo, phone interview with author, September 22, 2011.

155 *the anatomical and Mendelian paradigms* Douglas C. Wallace, "The Human Mitochondrion and Pathophysiology of Aging and Age-Related disease, in *Molecular Biology of Aging*, ed. Leonard Guarente et al. (New York: Cold Spring Harbor, 2008), 1.

CHAPTER 13

156 *the last three years of his marriage* Cynthia Kenyon, interview with author, December 29, 2009.

159 *to name a few such labs* Richard Miller, phone interview with author, December 13, 2011.

159 *food, stress, hunger, radiation* Anne Brunet, phone interview with author, November 16, 2011.

160 *when I entered the field* Andrew Dillin, interview with author, December 7, 2010.

160 *oh, isn't that interesting* Cynthia Kenyon, interview with author, December 29, 2009.

160 *molecular geneticists did not believe it* Tom Johnson, phone interview with author, January 27, 2012.

161 *insights were so new and varied* Cynthia Kenyon, phone interview with author, December 14, 2011.

163 *resveratrol made the cover* David Sinclair, et al., "SIRT1 is Required for AMPK Activation and the Beneficial Effects of Resveratrol on Mitochondrial Function," *Cell Metabolism* 15, no. 5 (2012): 675–90.

163 *first they say you're wrong* Leonard Hayflick, phone interview with author, November 8, 2011.

163 *she deserves credit* Nir Barzilai, phone interview with author, February 27, 2012.

164 *complex and promising metabolic regulators* Anne Brunet, phone interview with author, November 16, 2011.

164 *rapamycin, also known as sirolimus* "Rapamycin," accessed July 6, 2012, http://www.clinicaltrials.gov/ct2/results?term=rapamycin.

164 *resveratrol is under numerous clinical studies* "Resveratrol," accessed May 20, 2012, http://www.clinicaltrials.gov.

165 *what a drug company would do* Richard Miller, phone interview with author, December 13, 2010.

166 *a profound transition point* David Gems, phone interview with author, June 9, 2010.

166 *it was high risk, high reward* David Sinclair, phone interview with author, November 20, 2011.

166 *a drug to slow human aging* Matt Kaeberlein, phone interview with author, December 21, 2011.

167 *don't want to be the first to take the drug* Derek Lowe, phone interview with author, December 16, 2011.

167 *followed the first steps* Thomas Kuhn, *The Structure of Scientific Revolutions*, 3rd ed. (Chicago: University of Chicago Press, 1996).

167 *polished and well-spoken* Rozalyn Anderson, phone interview with author, December 18, 2009.

169 *is going to be amazing* Brian Kennedy, phone interview with author, March 15, 2011.

About the Author

Ted Anton is the author of *Bold Science: Seven Scientists Who Are Changing Our World* (W. H. Freeman, 2000), which was an Amazon.com Science Book pick and a featured choice on howthingswork.com. The *San Francisco Chronicle* described it as being "on the pioneering edge of science writing, spreading the notion that is a viable field of literary work."

His book, *Eros, Magic and the Murder of Professor Culianu* (Northwestern University Press, 1996), describing the unsolved murder of a University of Chicago professor, won the Carl Sandburg Award from the Friends of the Chicago Public Library. A finalist for a Book Award from the Investigative Reporters and Editors, it was reviewed in the *New York Review of Books, New York Times, Chicago Sun-Times and Tribune,* and *Washington Post,* as well as on National Public Radio and NBC. It has been published in German, Spanish, Italian, and Romanian. The *New York Times* called it "an engrossing story of a twentieth century original," and *Publishers Weekly* (starred review) praised it as "an intellectual thriller."

He was coeditor of the *New Science Journalists* (Ballantine, 1995). His articles for publications like *Lingua Franca, The Sciences, Chicago, Chicago Tribune, Publishers Weekly,* and others have been finalists for a National Magazine Award and have been included three straight years in *Best American Essays.*

A National Science Foundation grants reviewer, MacArthur Fellowship reviewer, and former Fulbright Senior Research Fellow, he speaks on longevity science and creativity on NBC, National Public Radio, WGN, and for government, industry, university, and high school groups at venues such as the New York Public Library, Newberry Library, the Harvard Club, University of Chicago Co-Op Bookstore, the American Cultural Center in Bucharest, and the American Association for the Advancement of Science's annual conference. His essay "Riff" appeared in *One Word: Contemporary Writers on the Words They Love and Loathe,* edited by Molly McQuade (Sarabande Books, 2010).

He is a professor of nonfiction writing at DePaul University in Chicago and is former chair of the Age Studies Executive Committee of the Modern Language Association.

Index

AARP, 141–42
acetyl, 30
adenosine monophosphate-activated protein kinase. *See* AMP-kinase
age discrimination, 146
age-1, 12, 20, 21, 25, 36, 38, 78
aging, xii; as catabolic, 90; definitions and meanings of, 6, 111; vs. development, 90; epigenetics of, 123–26; global, 140–46, 149, 151, 153–55 (*see also* aging battlefield; aging crisis); vs. life span/longevity, 6; as paradise, 148–51 (*see also* longevity dividend); theories of causation, 4; in the wild, 91–94. *See also specific topics*
aging battlefield, 145–48
aging crisis, xii–xiii, 55, 68–69, 111, 151–52. *See also* aging battlefield; aging: global
Alcedo, Joy, 93–94
AMPK (AMP-activated protein kinase), 69
AMPK1, 165
AMP-kinase (AMP-activated protein kinase), 69, 75, 105, 117, 134, 159
Anderson, Rozalyn, 58, 61, 78, 160, 163, 167, 171
antagonistic pleiotropy hypothesis, 3, 6
antioxidants. *See* free radicals; oxidative stress
Apfeld, Javier, 40–42
Arantes-Oliveira, Nuno, 67
Arking, Robert, 73, 81, 152
Ashkenazi Jews, 67, 81, 95, 105, 129–31
Aune, Roxanne, 145
Austad, Steven, 54–56, 73, 91; on aging, 77, 88; on aging crisis, 55; on *daf* and *sir* genes, 74; Richard Miller and, 78, 92; wild mice research, 78, 92
Austriaco, Nic, 23–25, 27
Ausubel, Fred, 34
Axinn, Donald, 53

Baker, Darren, 122
baker's yeast, 23–24. *See also* yeast
Barker, David, 94
Barnes, Barry, 133
Bartke, Andrzéj, 47–49, 74, 161
Barzilai, Nir, 96, 129–31, 133; at Albert Einstein College of Medicine's Aging Clinical Research Center, 135; Albert Einstein College of Medicine's Institute of Aging Research and, 129–32; background and early life, 128–29; centenarians and, 95, 105; CETP and, 132; Cobhar and, 137–38; on Cynthia Kenyon, 163; FOXO and, 95–96, 105, 131–32; heparin and, 136; insulin growth factor receptor and, 95; Longevity Project, 135; Pinchas Cohen and, 132, 138; where he is now, 171–72
Bayley, Cindy, 33, 42, 44, 52
bioethics, 54

Biomol, 58, 61, 62, 64, 65
Blair, Tony, 42–43
Bloom, David, 145
Brenner, Sydney, 7–8, 89–90
Brinkley, John, 4
Brown-Borg, Holly, 125, 172
Brunet, Anne, 105, 126, 136, 159, 164; background and overview, 75, 116–17; FOXO and, 75, 105, 159; insulin pathways and, 126, 136; laboratory, 75; where she is now, 172
Buck Institute, 52–53
Buffenstein, Rochelle, 93, 120–21, 172
Bush, George W., 54
Butler, Robert N., 76

Caenorhabditis elegans, xi–xii, 7, 9, 18, 50, 161, 169
caloric restriction, 47, 82; AMP-kinase and, 75; discovery of the benefits of, 5; early writings on, 5; FOXO and, 75; genes and, 58, 70, 71, 78, 107, 162, 163; insulin pathway and, 74; mechanism/pathway underlying the life-extending effects of, 31, 58, 70, 74, 81, 83, 107; rapamycin and, 81; research on, 10, 37, 45, 78, 81, 92, 106, 126, 159–60; resveratrol and, 64, 71, 83; sirtuins and, 58, 64, 70, 71, 78, 97, 162, 163
Calorie Restriction Society International, 5
Campisi, Judith, 45; Archventures and, 165; cancer and, 32, 45, 114, 121; on evolution, 31; inflammation and, 32, 114; Mayo Clinic and, 165; in the media, 64, 122–23; where she is now, 172
cancer, 32, 38, 60, 75, 114, 121–22; apoptosis and, 25; *daf-2* and, 74; DNA methylation and, 124, 125; epigenetics and, 123, 125–26; insulin and, 47, 106; Judy Campisi and, 32, 45, 114, 121; Leonard Guarente and, 24, 25, 58; metformin and, 119; mTOR gene and, 138; Nrf2 and, 121; p53 and, 58; rapamycin and, 79, 138, 164; resveratrol and, 110; stem cells and, 125; TOR and, 75, 80
cardiovascular health: resveratrol and, 96, 165; SIRT1 and, 126
cell division, 4
cell-stress resistance, 92, 93, 120–23, 156–59. *See also* oxidative stress
centenarians, 86, 95, 120, 129–31, 135; CETP and, 132, 135; characteristics of, 95, 132, 135, 144; FOXO and, 105, 132, 135, 138, 139, 164; genomes, 165; increasing number of, xiii, 55, 86, 128, 144; insulin and, 72, 95; longevity genes and, 39, 45, 60, 67, 72; Nir Barzilai and, 95, 105
CETP, 132, 135
Chang, Jeanne, 20–21, 40
cholesterol, 51
Clinton, Bill, 42–43
Cohen, Pinchas, 132, 138
Copernican revolution in biology, 166
cosmetic companies, 76–77, 106. *See also* Estée Lauder
Critser, Greg, 47
Cuervo, Ana Maria, 83
Curran, Sean, 37

daf, 74. See also *FOXO*
daf-2 ("Grim Reaper"), 20–21, 31, 34–41, 46, 67, 74, 78
daf-16 ("Sweet Sixteen"), 21, 34, 36–39, 42, 50, 67, 105, 161. See also *FOXO3a*
daf-23, 36
Darwin, Charles, 3
de Cabo, Rafael, 127
de Grey, Aubrey, 152–53
demography, is not destiny, 151–53
development: vs. aging, 90; as anabolic, 90
developmental revolution, xii, 89
Dhabi, Abu, 149
Dillin, Andrew, 50, 84, 105, 119, 160, 172; Cynthia Kenyon and, 22, 53, 115; FOXO and, 115; insulin and, 50, 105,

115; laboratory, 106, 114, 116, 117; in the media, 115, 116; reflections, 106; Rick Morimoto and, 136; on *SIR*, 116; where he is now, 172
Dipp, Michelle, 70, 82–83, 97, 102, 110
Distefano, Peter, 44, 102, 108
DNA, 133–34; damage to, 18, 123, 126; histones and, 30, 123; and longevity, 9; as "Midas's gold," 13; mitochondrial and nuclear, 82, 132; nature of, 8, 9; repair of, 126, 132; research on, 36, 39
DNA circles, ribosomal, 27–28
DNA methylation, 95, 124, 125
Dorman, Jenny, 38
Driscoll, Monica, 74, 77, 83–84, 159, 172
drugs: aging, 164–66; making, 118–20
Dupre, John, 133
Dyson, Freeman, 8

Easter Island, 79–81
elephants, 91
Elixir, 46, 60, 61, 70, 83, 98, 162–64; Cynthia Kenyon and, 44, 60; finances, 79, 98; ghrelin and, 69, 79, 83, 98; Leonard Guarente and, 44, 46, 60, 70, 77, 79, 98; metformin and, 69, 74; origins, 44, 46; resveratrol and, 70, 71
Ellison, Larry, 53
Encyclopedia of DNA Elements (ENCODE), 134
epigenetics, 123–26
Estée Lauder, 77, 102, 143
ethics of longevity research, 54
evolutionary theory, 160

Finch, Caleb, 13
Fishman, Ted, 146
Food and Drug Administration (FDA), 46, 51, 96–98, 119, 126
FOXO, 81, 95–96, 106, 115; Anne Brunet and, 75, 105, 159; caloric restriction and, 75; centenarians and, 105, 132, 135, 138, 139, 164; Nir Barzilai and, 95–96, 105, 131–32. See also *daf-16*
"*FOXO* hunters," 75, 95
FOXO transcription factor, 75, 93, 131–32, 138
FOXO3a, 95, 105, 132, 159
free radical theory of aging, 4, 160
free radicals, 81–82, 132, 135
Freud, Sigmund, 4
Friedman, David, 11, 12

Gems, David: on Cynthia Kenyon, 22, 33, 166; Linda Partridge and, 107–8, 134; resveratrol and, 78; *SIR* genes and, 107, 108, 110, 134; where he is now, 172–73
gene revolt, 9–11
gene trap. *See* promoter trap
GeneChip, 76
general electrophoresis, 36
genes: and aging, xiii, 10–13; "celebrity," 74 (see also *daf*; *SIR*); longevity, 117–18; silencing, 31. *See also specific topics*
genetic determinism, 133
genetic mutation accumulation theory, 4
genetic pathways, 75–77. *See also specific pathways*
genetics, molecular, 17–18
genetics revolution, 44
genome: human, 19, 27, 31, 34, 42–43, 75, 88, 129, 131, 133–35, 162, 163, 165, 167; mouse, 55–56
genome "book of life," 42–43
genomics, new, 117
Geron, 29, 45
"gerotech," 45
gerotechnology, 45, 119, 167
ghrelin, 69, 79, 83, 98
GlaxoSmithKline (GSK), 96–97, 103, 111, 164, 165; background and overview, 100; Sirtris and, 97, 100–1, 108–10, 164

Glenn, Paul, 103
Goldschmidt, Richard: ridiculed for believing in "hopeful monsters," 89
Gompertz, Benjamin, 148–49
"grandmother theory," 91–92
Greer, Eric, 159
Guarente, Leonard, 23–24, 27–32, 51–53, 103; on aging, 24; background and overview, 23–26; Brian Kennedy and, 23–26, 28, 71; cancer and, 24, 25, 58; Cynthia Kenyon and, ix, 26–27, 32, 42, 52, 79; David Sinclair and, 28, 53, 58–60, 70, 96; Elixer and, 44, 46, 60, 70, 77, 79, 98; Estée Lauder and, 77; Heidi Tissenbaum and, 42; laboratory, 27–29, 42, 53, 60, 70, 108; in the media, 31, 70–71, 96; Michael Rose on, 31; NAD and, 30, 53, 70; Nicholas Wade on, 31; Phillip Sharp and, 27; publications, 32; reflections, 25, 29–30; Nic Austriaco and, 23–25, 28; Shin Imai and, 29–30, 58; *SIR* genes and, 29–30, 45, 58, 70, 127; Sirtris and, 97; where he is now, 173

Hansen, Malene, 52
Harris, Jeanne, 33
Hart, Ron, 6
Harvard Life Span conference, 106
Harvey, Paul H., 21
Hayflick, Leonard, 4, 6, 88, 90, 103, 121, 163
Healthy Life Span conference, 102–3
heat shock proteins, 115–16, 121, 135
heat stress, 93, 115–16
Helfand, Stephen, 52, 111
Herndon, Laura, 74
hibernation and aging, 20–21, 32–36. See also *daf-2*
Hirsh, David, 9–10
histones, 30, 123–24
Holzenberger, Martin, 47
hormesis theory of aging, 4, 64
hormones, youth, 3
Horvitz, Robert, 8
Howitz, Konrad T., 58, 62–65, 68, 101
Hsin, Honor, 40–42
human clock, 94–96
Human Genome Project, 43, 55

Imai, Shin, 29–30, 126
inflammation, 114, 122
information revolution, 44
INK-ATTAC, 122
insulin, 35–37, 49; cancer and, 47, 106; centenarians and, 72, 95
insulin genes, 38, 39, 47, 50, 67, 72, 81. *See also* insulin growth factor
insulin growth factor, 37, 45, 49, 92, 105, 106, 158, 159
insulin growth factor 1 (IGF-1), 48, 95, 131, 161
insulin growth factor 2 (IGF-2), 95, 124
insulin growth factor receptor, 36, 45, 47, 95, 161
insulin growth factor signaling, 158, 159
insulin resistance, 47, 118, 122. *See also* insulin sensitivity
insulin sensitivity, 49, 69, 95, 105, 131, 134, 135; caloric restriction and, 37; ghrelin and, 79; resveratrol and, 165; stress resistance and, 92, 93; as universal element in healthy aging, 95. *See also* insulin resistance
insulin signaling, 38, 39, 41, 42, 49–50, 68, 95, 132. *See also* metformin
Issa, Jean-Pierre, 125

Jacoby, Susan, 147
Jazwinski, Michael, 24
Jews, Ashkenazi, 67, 81, 95, 105, 129–31
Johnson, F. Bradley, 28, 29, 126, 173
Johnson, Tom, 13, 160–61; *age-1* gene and, 20, 21, 25; background and overview, 11–13; Cynthia Kenyon and, 14, 19, 36; *daf-2*, 33, 36–37, 39; David Friedman and, 12; in the media, 12; Michael Klass and, 10–13; reflec-

tions, 11–13; shunned by scientific community, 13, 33; where he is now, 173
Judson, Horace, 13

Kaeberlein, Matt, 70, 111, 134; Andy Dillin and, 115; Brian Kennedy and, 70, 71, 76, 78, 80, 108; caloric restriction and, 70, 71, 108; on drugs to slow aging, 166–67; *Nature* and, 70–71, 107; resveratrol and, 70, 71, 78; *SIR* genes and, 70, 71, 108; TOR and, 76; where he is now, 173; yeast and, 70, 71, 78, 80, 108
Kass, Leon, 54, 55
Kennedy, Brian, 23–28, 83; in China, 111; on the future, 169; Kristan Steff and, 116; Matt Kaeberlein and, 70, 71, 76, 78, 80, 108; on media coverage as "runaway train," 102; resveratrol and, 78; *SIR* genes and, 71, 108; TOR and, 76; where he is now, 173; yeast and, 71, 78, 80, 108
Kenyon, Cynthia Jane, ix, xi–xiii, 14–22, 38–42, 46, 50, 52, 63, 90, 125, 156; Anne Brunet and, 75; background and early life, xi, 15–17; began testing compounds in human cells, 140; *C. elegans* and, xi–xii, 14; *daf-2* and, 74; *daf-16* and, 161; David Gems on, 166; David Reznick on, 99; David Sinclair and, 59, 63; developmental revolution and, 89; as director of Hilblom Center for the Biology of Aging, 55; Elixir and, 44, 60; finances, 55, 79; Gary Ruvkun and, 35; on genome and path to longevity, 75; impact, 33; insulin and, 44–45, 95; on insulin genes, 38, 60, 67; Jim Thomas and, 36; Judith Campisi and, 121; King Faisal International Prize, 55; laboratory, 38, 44, 50, 66–67, 106, 115; Linda Partridge on, 33; longevity pathway, 44–45; Nuno Arantes-Oliveira and, 67; originality and playful imagination, 166; Patrick O'Farrell and, 14–15; personality, xi, 14, 33–34, 53–54; Peter Thiel's funding of her research, 165; Ponce d'elegans, 121; publications, 26–27, 32, 50, 59; reflections, 93, 139, 156, 160, 161; on resveratrol, 70; on science and scientists, 67, 70, 93, 102, 109–10; speeches, 139; Tom Johnson and, 12, 14, 36; on transcription factor, 81; vision, 163; where she is now, 175
Kenyon, Jane, 15–17
kinship dynamics, 91
Kirkwood, Tom, 6
Kitt, Eartha, 31
Klass, Michael, 9–13, 24, 173
Koumaru, Koutaru, 36
Kuhn, Thomas S., 34, 71–72, 167
Kurzweil, Ray, 152–53

Lederberg, Joshua, 8
Lewis, Thomas, 82
Libina, Natasha, 67
life span: increasing, xiii, 143–45 (*see also* aging battlefield; aging crisis; aging: global). *See also* centenarians; *specific topics*
LifeGen, 45
lifetime, phases of, 73
Lithgow, Gordon, 78, 105, 114
longevity: ethical issues and criticism of research on, 54; genetics and, 6; nonprofit organizations, 141–42; paradigm shift in the biology of, 32; of various animals, 87–88
longevity companies, 45–46, 165. *See also specific companies*
longevity dividend, 91, 96, 149
longevity gene pitfalls, 133–35
longevity mechanisms, 135. *See also specific mechanisms*
"longevity mutants," human, 95
longevity revolution, 142, 150–51. *See also specific topics*

Longo, Valter, 81, 106, 138, 155, 173–74
Lowe, Derek, 97, 109, 111, 167

MacArthur Foundation Research Network on an Aging Society, 142
Maes, Daniel, 77
Malthus, Thomas, 150
mammalian target of rapamycin (mTOR), 80–81, 105, 116, 137, 138, 159
Martin, George, 13, 22, 52
McCabe, Edward, 90
McCabe, Linda, 90
McCay, Clive, 5
media coverage, 142; becoming a "runaway train," 102–4. *See also specific topics*
MELAS syndrome, 83
Mencken, H. L., 4
metformin, 69, 74, 119, 164
Miller, Richard, 51–52; on Alzheimer's disease, 49; cell-stress resistance and, 159; on insulin genes and pathways, 50; Jan Vijg on, 103; mice research and, 48, 78, 92; NIA and, 49, 165; Steven Austad and, 78, 92; where he is now, 174
Mills, Kevin, 60
mitochondria, 81–83, 132, 135, 155; resveratrol and, 82, 101, 108, 163
mole rat, common, 120–21
Molecular Biology of the Gene (Watson), 17–18
Molecular Genetics of Aging, 52
Morimoto, Rick, 74, 93, 105, 116, 119, 136–37, 159
mosaics (genetics), 40, 41
Moss, Cynthia, 91
mouse genome, 55–56
mouse research, 47–50, 55–56, 80, 83, 92, 127. *See also specific topics*
MPM Capital, 60, 79, 98
Murphy, Coleen, 50, 66–67, 137, 139, 159, 174
mutation accumulation, 4
myelodysplastics syndrome (MDS), 125–26

NAD (nicotinamide adenine dinucleotide), 53, 70, 78
National Institute on Aging (NIA), 5–6, 35, 160, 165
National Institutes of Health (NIH), 49, 76
Neugarten, Bernice, 148, 150
nicotinamide adenine dinucleotide (NAD), 53, 70, 78

obesity, 79, 81; resveratrol and, 83, 96, 97, 101–3, 127, 134, 138, 162, 163, 165
O'Farrell, Patrick, 14–15
Ogg, Scott, 38
Okinawa Centenarian Study, 131–32
Olshansky, S. Jay, 144, 148–50, 153; background and overview, 54; Leon Kass and, 54, 55; skepticism and cynicism, 54–56, 96; where he is now, 174
Oprah Winfrey Show, The (TV program), 101, 104, 156
oxidative stress, 47, 75, 93, 121, 132, 155, 162. *See also* cell-stress resistance; free radicals
Oz, Mehmet, 101, 104

p53 gene, 58
paradigm shifts, 71–72
Partridge, Linda, 33, 67; on evolution of aging, 21; fly research, 49, 76, 78, 107–8; FOXO and, 76; insulin research, 49; resveratrol and, 78, 111; on scientific research, 111; *SIR* genes and, 107–8, 111, 134; where she is now, 174
PAX6, 89
Pfizer, 108–10
phenols. *See* polyphenols
PNC1, 53, 70; as "master regulator" of longevity, 57–58

polyphenols, 58, 64, 78, 97, 101. *See also* resveratrol
population aging. *See* aging battlefield; aging crisis; aging: global; centenarians: increasing number of
Presidential Commission for the Study of Bioethical Issues, 54
Prolla, Tomas, 33, 76, 106, 174–75
promoter trap, 15
protein folding, 93, 105, 115–16
Ptashne, Mark, 26
Puigserver, Pere, 82

Rando, Tom, 103, 126, 159
rapamycin, 80, 113, 134, 159, 164; and cancer, 79–80, 138, 164; history and overview, 76, 79–80; and longevity, 80, 81, 105. *See also* mammalian target of rapamycin; target of rapamycin
rDNA, 27–28
Rennie, Drummond, 46
reproduction: in elephants, 91; trade-off between longevity and, 39–41
resveratrol, 82–83, 102, 164–65; Brian Kennedy and, 78; caloric restriction and, 64, 71, 83; cancer and, 110; and cardiovascular function, 96, 165; challenges to the claims about, 102, 107, 108; Cynthia Kenyon and, 70; David Gems and, 78; David Sinclair and, 58–59, 61–66, 78, 101, 103–4, 108, 163; dosage, 78; Dr. Oz on, 104; Elixir and, 70, 71; food sources of, 61; forms and mimetics, 83, 108, 110, 119 (*see also* SRT compounds; SRT-501); insulin sensitivity and, 165; Johan Auwerx and, 160; Linda Partridge and, 78, 111; and longevity, 63, 78, 97, 102–4, 127, 162; Marc Tatar and, 63; Matt Kaeberlein and, 70, 71, 78; mechanism of action, 82; in the media, 63–64, 104, 163; and mitochondria, 82, 101, 108, 163; nature and overview of, 61–62; NIA and, 164–65; obesity and, 83, 96, 97, 101–3, 127, 134, 138, 162, 163, 165; patentability and efforts to patent, 61, 65, 83, 96; and the power of an idea, 110–12; products containing, 102; public hunger for, 104, 110; *SIR2* and, 71; SIRT1 and, 83, 108, 110, 119, 162, 163, 165; Sirtris and, 101–2, 110; sirtuins and, 64, 66, 97, 107, 108, 119; yeast and, 59, 61, 64, 70, 78
Reznick, David, 87, 88, 91, 93, 99, 125
Riddle, Don, 37
RNA interference (RNAi), 50
Roizman, Bernard, 98
Rose, Michael, 24, 31
Ruvkun, Gary, 35, 37, 52–53, 125; background, 34–35; cloning method, 38; Cynthia Kenyon and, 22, 34–36, 38–39; on *daf* and *SIR* genes, 74; *daf-2* and, 34, 35; *daf-2* insulin paper, 39; *daf-16* and, 34, 161; *daf-23* and, 36; David Sinclair and, 60, 63; developmental revolution and, 89; on evolutionary biologists, 40; FOXO and, 106; on HOX gene discovery, 89; insulin and, 95; laboratory, 34, 35, 38, 39, 50, 106, 161; Lasker Award, 106; in the media, 34, 36, 37, 50, 161; RNA interference and, 50; *Science* paper, 38–39; where he is now, 175

savannas, east African, 93
Science of Aging Knowledge Environment (SAGEKE), 45
scientific revolutions, 34, 71–72, 166, 167
Scolnick, Ed, 60
sea birds, 92
sex. *See* reproduction
Shaklee, 103–4
Shapiro, Ben, 60
Sharp, Phillip, 27, 68
signaling pathways, 75, 105, 117, 158–59. *See also specific pathways*

Sinclair, David, 28, 72; in Australia, 59; background and overview, 28, 57, 59, 60; Biomol and, 58–60, 62, 64; Christoph Westphal and, 65–68; Cynthia Kenyon and, 59, 63; Derek Lowe and, 109; Elixir and, 70; Estée Lauder and, 77, 102; finances, 65, 68, 77, 96, 101; Gary Ruvkun and, 60, 63; Konrad Howitz and, 58, 62–64, 68; Kuhn's *Structure of Scientific Revolutions* and, 71–72; laboratory, 57, 60–62, 78; Leonard Guarente and, 28, 53, 58–60, 70, 96; Marc Tatar and, 62, 63, 78; Matt Kaeberlein and, 71; in the media, 63–65, 77, 83, 96, 97, 103, 104, 111; personality, 28; *PNC1* gene and, 53, 70; reflections, 166; resveratrol and, 58–59, 61–66, 78, 101, 103–4, 108, 163; rudeness, 109, 110; Shaklee and, 103–4; *sir* gene and, 70; *SIRT1* gene and, 58, 64, 163; Sirtris and, 68, 96, 101; sirtuin and, 58, 64, 101, 126, 163; where he is now, 175; xenohormesis hypothesis, 64; yeast cell research, 28, 57, 64, 70

SIR (silent information regulator), 27, 80

Sir2, 30, 58, 71

SIR2, 29, 30, 42, 58, 71, 77, 78, 107, 127

sir-2.1, 42, 61, 107–8, 134

sirolimus. *See* rapamycin

SIRT1, 58, 119, 134, 137; and cardiovascular disease, 126; resveratrol and, 83, 108, 110, 119, 162, 163, 165

SIRT6, 126

Sirtris Pharmaceuticals, 65, 96–98, 106, 108, 110, 118–19, 134, 137, 162, 164; challenges faced by, 70, 79; Christoph Westphal and, 68, 83, 85, 96, 100, 104, 106, 108, 110; David Sinclair and, 68, 96, 101; Derek Lowe on, 109; finances, 68, 70, 79, 83, 85, 101; GlaxoSmithKline and, 97, 100, 101, 108, 164; Healthy Life Span conference, 102–3; vs. Healthy Lifespan Institute, 110; incorporation of, 68; Michelle Dipp and, 82–83, 97, 102; resveratrol and, 83, 96, 97, 101–2, 104, 110

sirtuins, 158–59; caloric restriction and, 58, 64, 70, 71, 78, 97, 162, 163; David Sinclair and, 58, 64, 101, 126, 163; resveratrol and, 64, 66, 97, 107, 108, 119. *See also specific topics*

skin-care products, 77, 106. *See also* Estée Lauder

Slagboom, P. Eline, 33, 94–95, 132, 134, 135, 175

Slaoui, Moncef, 96, 100–1

Sonneborn, Joan Smith, 9

SRT compounds, 96, 108, 109, 118–19

SRT501, 83, 96, 101, 110

Stahl, Leslie, 12

statin drugs, 51

Steffen, Kristan, 116

Steinach, Eugen, 4

Stipp, David, 83, 97

stress resistance. *See* cell-stress resistance

Sulston, John, 8

surgeries purported to increase longevity, 4

Suzuki, Makoto, 131

Tabtiang, Ramon, 19, 20

target of rapamycin (TOR), 75, 76, 80, 81, 116. *See also* mammalian target of rapamycin

Tatar, Marc, 63, 67, 74–75, 77, 91, 103; background and overview, 62; David Sinclair and, 62, 63, 78; on insulin, 49; resveratrol and, 63; where he is now, 175

telomerase gene, 29

telomere, 18, 29, 93

telomere lengthening, 29, 92, 105

Thomas, Jim, 36, 50, 67

Tissenbaum, Heidi A., 35, 36, 39, 42, 53, 113, 175

trade-off theory of aging. *See* antagonistic pleiotropy hypothesis
transcription factors, 39, 81, 161. See also *FOXO*
tumor suppressor mechanisms, 121–22

Van Deursen, Jan M., 122
Varmus, Harold, 71
vasectomies, purported to increase longevity, 4
Vaupel, James, 153
Vijg, Jan, 103
Vlasuk, George, 104, 110, 119, 137
Voisine, Cindy, 136

Wade, Nicholas, 31, 64, 102, 104, 106, 123
Walford, Roy, 5
Walker, Graham, 18
Wallace, Douglas, 155
Watson, James, 17–18
Weindruch, Rick, 33, 76, 106. *See also* LifeGen
Westphal, Christoph, 102, 110; Acceleron and, 66; background and overview, 65–66; David Sinclair and, 65–68; finances, 68, 83, 96, 100, 101; GlaxoSmithKline and, 96, 100, 101, 108; in the media, 104; personality, 65; Polaris Venture Partners and, 65, 68; Sirtris and, 68, 83, 96, 100, 104, 106, 108, 110
White, John, 7–8
Willcox, Bradley, 105, 131–35, 138
Willcox, Craig, 105, 131, 133
Williams, George, 4, 50–51
Winfrey, Oprah, 101, 104, 156
Witty, Andrew, 100–1, 109
Woese, Carl, 8
Woodward, Kathleen, 154
World War II, 94
worms: research on, xi–xii, 7–13, 20–22, 32–34, 36–43, 48, 67, 74, 76–78, 81, 106–8, 136–38, 159, 161. See also *Caenorhabditis elegans*
Wyeth, 80, 104

xenohormesis hypothesis, 64

yeast, 23–28, 32, 59, 70, 77–78, 81; resveratrol and, 59, 61, 64, 70, 78. *See also specific topics*
Youth (UTH 1–4), 27